COLOR ATLAS OF
HUMAN DISSECTION

C. C. Chumbley

Lecturer in Anatomy
St. George's Medical Hospital School
London

R.T. Hutchings

Freelance Photographer
Formerly Chief Medical Laboratory Scientific Officer
Royal College of Surgeons of England

Year Book Medical Publishers Inc

Copyright © C.C. Chumbley, R.T. Hutchings, 1988
Published by Wolfe Publishing Limited, 1988
Printed by W.S. Cowell Ltd, Ipswich, England
Typeset in Linotron Times New Roman and Italic

For a full list of other atlases please write to Year
Book Medical Publishers, Inc., 200 North LaSalle
Street, Chicago, Illinois 60601.

Library of Congress Cataloging-in-Publication Data

Chumbley, C.C.
    A color atlas of human dissection/C.C. Chumbley, R.T. Hutchings.
        p. cm.
    Includes index.
    ISBN 0-8151-1660-8
    1. Anatomy. Human—Atlases. 2. Human dissection—Atlases.
I. Hutchings. R.T. II. Title.
    [DNLM: 1. Dissection—atlases.   OS 17 C559c]
OM25.C478 1988
611'.0022—dc19
DNLM/DLC
for Library of Congress   88-17122
                                                    CIP

Preface

This book results from our belief that there is no substitute for dissection of a cadaver for a true understanding of human anatomy.

Medical courses are so crammed with information that students often come to the dissecting room unprepared for the session. The purpose of this book is to offer a scheme of practical anatomy, with a structure flexible enough to allow it to be used in a modern anatomy course.

The dissection procedures are arranged in such a way that the normal relationships between structures are preserved as far as possible. The specimen may be referred to again later, in conjunction with one or more of the standard textbooks of anatomy, and the dissected parts can be reassembled for a complete review of the body at the end of the course.

The instructions given throughout are of two kinds, those directing the work of dissection of a particular region, and those suggesting ways to study the specimen. It is usual that several students are assigned to one cadaver, and although only one member of the group may dissect, all students should take time to study the specimen (see annotated sample page below, explaining how best to use the book in this respect).

69

The Sole of the Foot

opening text, introducing the section

This is a difficult dissection. You will have to display delicate structures running through dense connective tissue. Use a sharp scalpel and fine forceps, and be careful to clean each structure as it is uncovered in the course of dissection.

Remove the skin and superficial fascia from the sole of the foot and flexor surfaces of the toes, by starting at the heel and reflecting both layers forward to the toes (**1**). Avoid damaging the *plantar aponeurosis*.

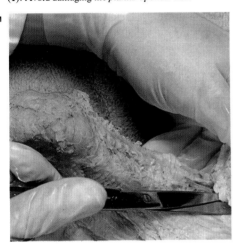

Clean the *plantar aponeurosis*.

Remove the plantar aponeurosis piecemeal to display the first layer of muscles of the sole (**2**).

Identify:
○ *the adductor hallucis,*
○ *the flexor digitorum brevis,*
○ *the adductor digiti minimi.*
Search for the medial and lateral plantar nerves and vessels. Establish the continuity of these nerves and arteries with the tibial nerve and posterior tibial artery, behind the medial malleolus and in the posterior compartment of the leg.

Separate the flexor digitorum brevis muscle from the underlying structures. Remove the belly of the muscle, but leave the tendons attached to the toes. Clean and identify the muscles of the second layer of the sole.

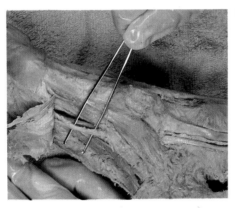

2

325C

page and figure reference to the Colour Atlas of Human Anatomy

Identify the tendon of flexor digitorum longus, flexor accessorius and the lumbricals. Note that the tendon of flexor hallucis longus crosses the sole deep to the tendon of flexor digitorum longus. Identify both long flexors in the posterior compartment of the leg.

Divide the tendon of flexor digitorum longus just where it is joined by the flexor accessorius. Reflect the anterior part forwards, to display muscles of the third layer of the sole in the anterior half of the foot. Clean these structures.

instructions for dissection

Identify the flexor hallucis brevis, flexor digiti minimi and the two parts of adductor hallucis. Cut through the origin of the flexor hallucis brevis, and reflect the muscle forwards to display the two sesamoid bones in its tendons below the head of the first metatarsal.

Divide the oblique head of adductor hallucis, and reflect the muscle to display the plantar arterial arch and deep branch of the lateral plantar nerve. Clean the plantar arterial arch and determine its formation and branches.

instructions to be followed by all students as they study the dissected specimen

The fourth layer consists of the interossei. Force the metatarsals apart, to observe their attachments.

For a thorough study of the completed dissection, the student will require one of the many excellent atlases of gross anatomy currently available. In this manual, we make constant reference to the second edition of 'A Colour Atlas of Human Anatomy' by R.M.H. McMinn and R.T. Hutchings, and have consistently followed the order of anatomical regions adopted in that atlas. Since teachers of anatomy may prefer to approach dissection in a different order, alternative schemes are offered in the appendix. A section on techniques of dissection discusses and illustrates the fundamental steps involved in every procedure, and draws the student's attention to the importance of treating the cadaver with care.

Our work concentrates on guiding the student towards an appreciation of the topography of the human body in three dimensions. We have omitted dissections that are too difficult and discouraging to the student, and also a great deal of the customary fine detail of anatomy.

We have used colour photographs to bring students as close as possible to the reality of the dissecting room. The majority of the photographs represent a true sequence of dissection of the same male and female cadavers. The photographic display in two dimensions of the complex three-dimensional structures of the human body can be fraught with problems, many of which have been avoided by adopting some standardized photographic techniques. Where possible, viewing angles, lighting and reproduction ratios were kept within predetermined limits. Perspective control was employed with the aid of short telephoto lenses and an 'all movements' camera, in an attempt to eliminate the 'monocular' view usually associated with the ubiquitous 35mm camera. Film type, filtration and processing have been consistent for all pictures, surface anatomy and osteology included, to avoid any vagaries which can easily arise in this type of work.

We are both grateful to the staff and students of St. George's Hospital Medical School. Thanks are due to Professor P.N. Dilly for providing us with facilities in his department. The encouragement we received from students and demonstrators while testing the earlier drafts was extemely valuable; in particular, Dr H.Y. Charlie Chan and Dr Andrew Yelland very kindly gave us their time and expertise. We are grateful for the unstinting assistance provided by Mr Frank Simpson and Mr Charles Josling at all stages of the preparation of the book.

At Wolfe Publishing, we thank Patrick Daly for suggesting that we collaborate on this work, as well as for his guidance and encouragement during its earlier stages. We extend our special thanks to Helen Hadjidimitriadou for her tireless work in guiding this book to publication.

C.C.C.
R.T.H.

Certain illustrations used in this book have been borrowed from other Wolfe Colour Atlases. Details are as follows:

1. From **A Colour Atlas of Surface Anatomy**, K.M. Backhouse, R.T. Hutchings:
   Section 19: Figs 2,3,4.
   Section 27: Fig. 1.
   Section 36: Figs 2,3,4,5,6,7.
   Section 42: Fig. 2.
   Section 44: Fig. 1.

2. From **A Colour Atlas of Optic Disk Abnormalities**, E.E. Kritzinger, H.M. Beaumont:
   Section 19: Fig. 5.

3. From **A Colour Atlas of Ear Disease**, R.A. Chole:
   Section 19: Fig. 6.

4. From **A Colour Atlas of Fibreoptic Endoscopy of the Upper Respiratory Tract**, J.D. Shaw, J.M. Lancer:
   Section 19: Figs 7a,7b,10.

Contents

Techniques of Dissection

Care of the cadaver
In most countries, donors will have bequeathed their bodies to medical schools for dissection, in order to contribute in the medical students' learning of anatomy. It is a very special privilege to be allowed to dissect a human body, and the cadaver which has been allocated to you should be treated with respect.

To treat a cadaver with respect is firstly to dissect it with great care so that the maximum benefit is obtained from the dissection and, secondly, to preserve it carefully so that it does not dry out or decay during the time it is in your care.

- Keep the cadaver covered with moistened towelling or muslin and plastic sheeting between dissecting sessions.

- Uncover only as much of the body as is necessary for the particular dissection or study session.

- Wash out the dissection occasionally.

- Keep the area you are currently dissecting or studying moist, by using the wetting solution provided by the department of anatomy.

- Report any damage, by mould for example, to the dissecting room staff, so that it is dealt with.

In most countries, there are laws and regulations controlling the treatment of human bodies while in the care of a department of anatomy for dissection. Ensure that you understand these legal requirements so far as they affect you.

Dissecting instruments
Each group of students should have a set of dissecting instruments consisting of (**1**):
- a pair of large scissors with blunt tips,
- a pair of small pointed scissors,
- a scalpel with disposable blades or a fixed blade,
- fine and blunt forceps,
- a strong, blunt probe,
- a skin pencil, for marking incision lines on the skin.

1

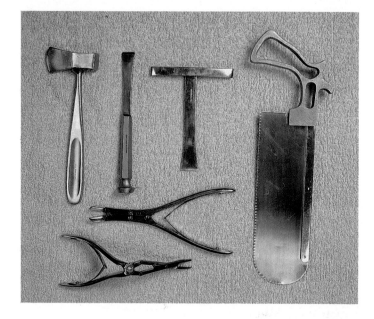

The department of anatomy will provide larger instruments which you will need occasionally (**2**):
- a saw,
- rongeurs (bone nibblers),
- bone cutting forceps,
- osteotomes (chisels) and a mallet,
- a vibrating saw.

2

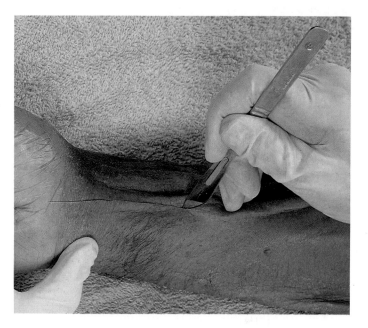

Making an incision in the skin
The thickness of the skin is only one or two millimetres. To avoid damaging the underlying structures, use the tip of the scalpel blade and rest the heel of your hand on the specimen (**3**).

3

Reflection of the skin
Using blunt forceps, hold a corner of the skin between two incisions, and with a sharp scalpel carefully divide the white fibres which bind the skin to the superficial fascia (**4**).

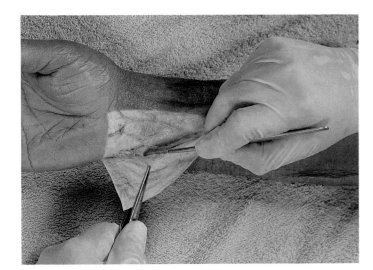

If a large area of skin is to be removed, cut a 'buttonhole' in the skin flap and hook your finger through it. This method is less tiring than holding the skin flap with forceps (5).

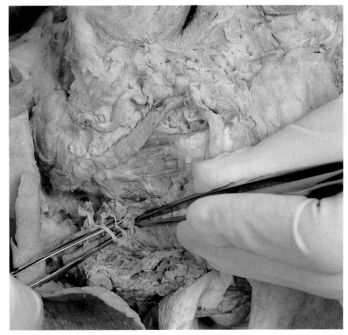

Reflection of the superficial fascia
Figure 6 shows the skin reflected and the superficial fascia in place. The fatty superficial fascia is best removed by blunt dissection, using fingers (7); if it is too dense for your fingers, use closed blunt forceps or closed scissors to break down the dense connective tissue (8).

Developing the fascial planes between muscles
Use your fingers, or a blunt probe, to break down the loose connective tissue between muscles (9 & 10). In this way, you will preserve the neurovascular bundles passing between the muscles.

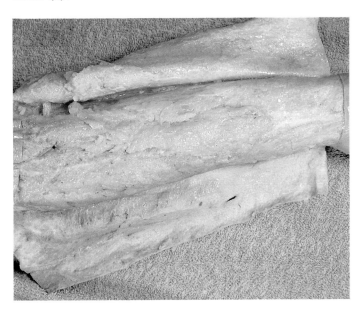

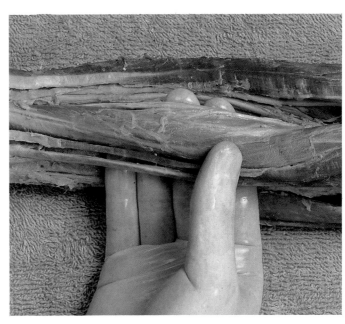

10

Cleaning muscles

This process involves the use of a sharp scalpel and fine forceps, to remove the connective tissue from the surface of a muscle. The areas of muscle that are necessary to clean are:

- the borders,
- the origin and insertion,
- enough of the body of a muscle, to see the general direction taken by its fibres (**11**).

11

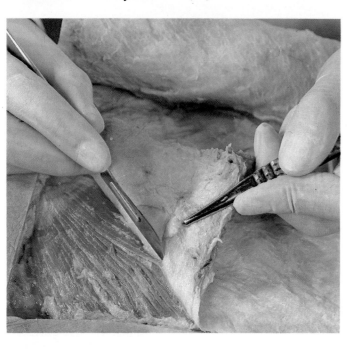

12

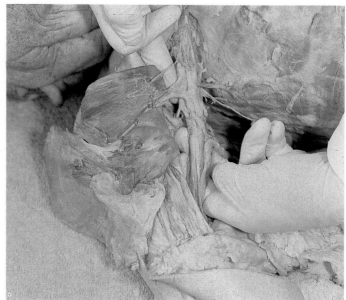

13

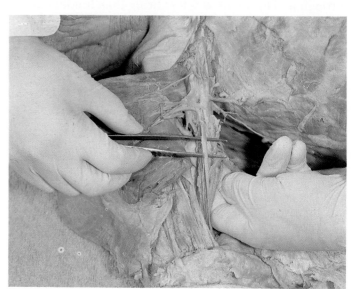

Separating the structures in a neurovascular bundle

This method can be used both with small and large vascular bundles.

First, define the bundle by developing the fascial planes around it with your fingers (**12**). Next, open the connective tissue sheath, using scissors if necessary. Finally, hold the neurovascular bundle with a pair of forceps, and use either another pair of forceps or scissors to separate the structures, by running the instrument between the structures parallel to the long axis of the bundle (**13**).

Preserve nerves and arteries, but you can sacrifice veins in order to make your dissection clearer.

Head and Neck

1

The Posterior Triangle of the Neck

In this dissection the skin is reflected from the front and sides of the neck. Then the borders of the posterior triangle are demonstrated, together with its nerves and vessels. The concept of the layers of cervical fascia is introduced.

Make three incisions on the front and sides of the neck (**1**):
- from the point of the chin to the *jugular notch*,
- along the length of the clavicles, and
- from the point of the chin along the margin of the mandible, from the *angle of the mandible* across the *mastoid process* to the *external occipital protuberance*.

Find the plane between the *platysma muscle* and the *investing layer of deep cervical fascia* (**2**). Reflect the skin and platysma laterally, starting at the angle between the vertical and clavicular incisions. As the flap is reflected past the inferior pole of the *parotid salivary gland*, just posterior to the angle of the mandible, try to locate the *cervical branch of the facial nerve*, which supplies the platysma.

40 A
Locate the sternocleidomastoid and trapezius muscles beneath the investing layer of the deep cervical fascia.

*With your finger, outline the **borders of the posterior triangle**: the posterior border of the sternocleidomastoid muscle, the middle third of the clavicle, the anterior border of the trapezius muscle.*

*Observe the **external jugular vein** as it passes anterior to the sternocleidomastoid muscle, and note where it pierces the investing layer of the deep cervical fascia.*

Make an incision in the investing layer of the deep fascia, taking care to avoid damage to the nerves you have found, and the external jugular vein. Cut along:
- the posterior border of the sternocleidomastoid muscle,
- the middle third of the clavicle, and
- the anterior border of the trapezius muscle.

Carefully remove the connective tissue, fat and lymph nodes from the posterior triangle, as far as the prevertebral fascia (**3**).

40 A
41 B
Identify and preserve:
- *the **accessory nerve**,*
- *the **branches of the cervical plexus**,*
- *the **inferior belly of the omohyoid muscle**,*
- *the **axillary sheath**, containing the **brachial plexus** and the **subclavian artery**,*
- *the **subclavian vein**.*

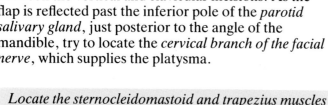

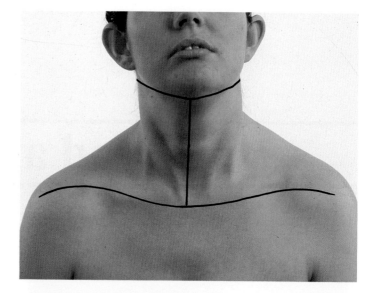

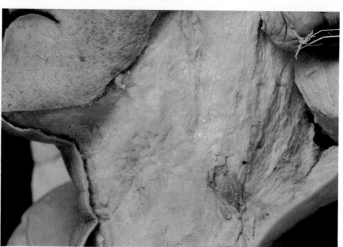

1

2

3

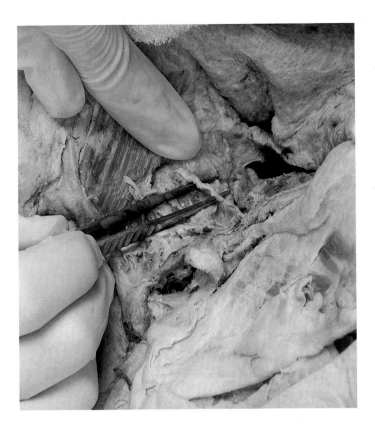

*Find the **cutaneous branches of the cervical plexus** as they radiate from the mid-point of the posterior border of the sternocleidomastoid muscle.*

*Find the position of the **accessory nerve**. Its course can be found by bisecting, at right angles, the line between the **mastoid process** and the **angle of the mandible** (4).*

Note the three layers of cervical fascia which have been demonstrated in this dissection:
- *the superficial fascia, beneath the skin which contains the platysma muscle,*
- *the investing layer of deep fascia, which forms the roof of the triangle and also encloses the sternocleidomastoid and trapezius muscles,*
- *the prevertebral fascia, which forms the floor of the triangle.*

Arrange the cutaneous branches of the cervical plexus in their usual positions and indicate the area of skin supplied by each of them. Correlate the root value of each nerve with the dermatomes in the area it supplies.

The muscles making up the floor of the posterior triangle will be studied in the dissection of the prevertebral region (Section 13).

2

The Anterior Triangle of the Neck and the structures deep to the sternocleidomastoid muscle

To facilitate the understanding of this complex region, the anterior triangle is subdivided into smaller regions and the contents of each of the subdivisions are demonstrated in turn. Finally, the sternocleidomastoid muscle is divided and the contents of the carotid sheath are displayed.

The skin and superficial fascia were reflected from the anterior triangle in the previous dissection.

- Place a block between the shoulders of the cadaver to extend the neck.
- Find the hyoid bone by palpation (**1**).
- Remove the deep fascia between the mandible and the hyoid bone, and between the hyoid bone and the anterior borders of the sternocleidomastoid muscles.
- Clean both bellies of the *digastric muscle*, and the *superior belly of the omohyoid muscle* (**2**).

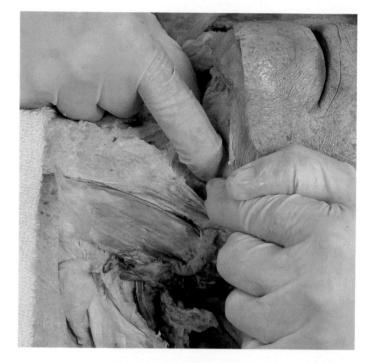

Define the borders of the anterior triangle:
- *the margin of the mandible,*
- *the anterior border of the sternocleidomastoid muscle, and*
- *the mid-line of the neck.*

41 B

Use the hyoid bone, digastric and omohyoid muscles to define the subdivisions of the anterior triangle:
- *the* **submandibular triangle**, *between the mandible and the digastric muscle,*
- *the* **muscular triangle**, *between the hyoid bone and the anterior margins of the superior belly of the omohyoid muscle,*
- *the* **carotid triangle**, *between the anterior border of the sternocleidomastoid muscle, the inferior border of the posterior belly of the digastric muscle, and the posterior border of the superior belly of the omohyoid muscle,*
- *the* **submental triangle**, *found between the anterior bellies of the digastric muscle and the hyoid bone.*

Examine the **muscular triangle**. *Find:*

41 B
- *the* **omohyoid** *and* **sternohyoid muscles** *superficially,*
- *the* **thyrohyoid** *and* **sternothyroid muscles** *at a deeper level.*

Push these muscles aside to find:

43 B
- *the* **larynx**,
- *the* **lobes** *and* **isthmus** *of the* **thyroid gland**.

Note that the thyroid gland is bound to the larynx and trachea by the **pretracheal fascia**.

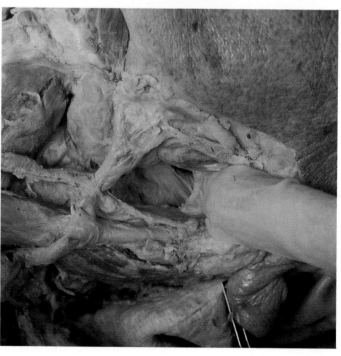

1

2

3

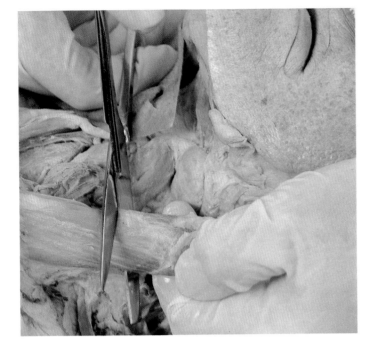

Identify the following structures in the midline of the neck from the chin to the jugular notch:
- *the **mandible**,*
- *the **mylohyoid muscle** in the floor of the submental triangle,*
- *the **body of the hyoid bone**,*
- *the **thyrohyoid membrane**,*
- *the **thyroid cartilage** of the larynx, with its notch and prominence,*
- *the **cricothyroid membrane**,*
- *the **cricoid cartilage** of the larynx,*
- *the rings of the **trachea**,*
- *the **isthmus of the thyroid gland**,*
- ***trachea** again,*
- *the **jugular notch** between the medial ends of the clavicles and the top of the sternum.*

Having found all these structures on dissection, locate them by palpation on the front of your own neck or the neck of a colleague.

4

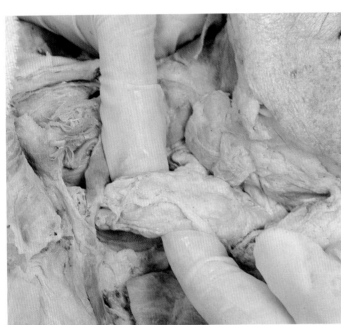

Examine the carotid triangle. Find the carotid sheath. Carefully remove the carotid sheath to display its contents.

Identify the following contents of the carotid triangle:
- *the **common**, **internal** and **external carotid arteries**,*   **41 B**
- *the origin, from the external carotid artery, of the **superior thyroid, lingual, facial** and **occipital arteries**,*
- *the **internal jugular vein**,*
- *the **hypoglossal nerve**, taking careful note of its relations,*
- *the **ansa cervicalis**, noting that its branches supply some muscles in the muscular triangle.*

Leave the detailed examination of the contents of the submandibular triangle for a later dissection.
- Clean the sternocleidomastoid muscle, and note its origin and insertions.
- Divide the sternocleidomastoid muscle just below the origins of the cutaneous branches of the cervical plexus (**3**). Reflect the divided muscle upwards and downwards.
- Clean the structures deep to the sternocleidomastoid muscle, by removing the remains of the deep fascia and carotid sheath (**4 & 5**).

5

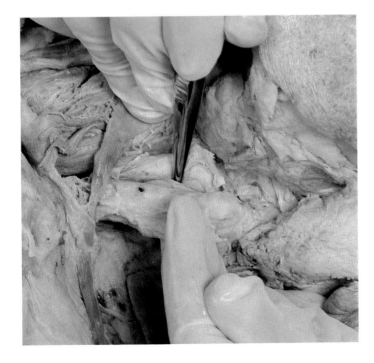

43 B

Identify:
○ the **common, internal** and **external carotid arteries**,
○ the **internal jugular vein** and some of its tributaries,
○ the larger **deep cervical lymph nodes**,
○ the **ansa cervicalis**, if it was not found in the carotid triangle,
○ the **intermediate tendon** of the **omohyoid muscle**,
○ the **prevertebral fascia**, covering the **scalenus anterior muscle** and forming the **axillary sheath** over the **brachial plexus** and **subclavian artery**.

Review the positions and attachments of the layers of deep cervical fascia, as they have been seen in this dissection and the previous one:
○ the investing layer,
○ the pretracheal fascia,
○ the carotid sheath,
○ the prevertebral fascia.

Insert them on the diagram below (**6**).

6

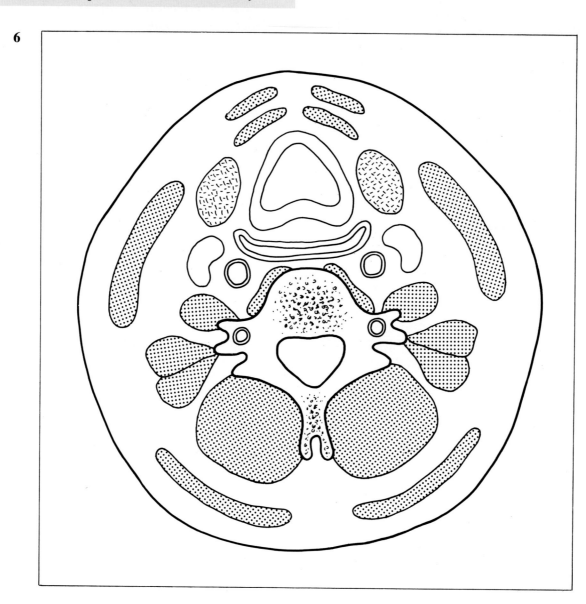

3

The Face

In this dissection the skin is reflected from the face so that the muscles of facial expression can be examined. The facial vessels and some cutaneous branches of the trigeminal nerve are demonstrated.

1

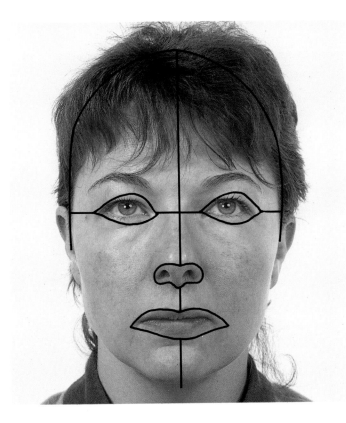

Make the following incisions in the skin of the face and scalp, taking great care to avoid damage to the underlying structures (**1**):
- from the vertex, to meet the incision which displayed the triangles of the neck, just in front of the auricle,
- from the vertex to the chin, encircling the nostrils and mouth as close as possible to the borders of the lips,
- from the *nasion*, encircling the orbits to meet the first incision just above and in front of the auricle.

Reflect the skin between the second and third incisions, again taking care to minimize damage to the underlying structures which are located in the superficial fascia (**2**). The facial muscles insert into the skin.

Clean the sphincter muscle of the lips, the *orbicularis oris*.

Clean the radial muscles around the mouth. They are often difficult to display convincingly, but you should be able to demonstrate most of them.

34 A

2

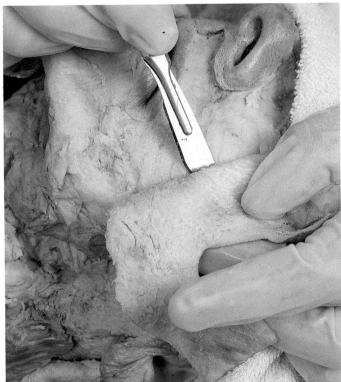

Divide the *levator labii superioris* close to its origin below the orbit, to display the *infraorbital branch of the maxillary nerve*.

Look deep to the *depressor anguli oris* to find the *mental branch of the mandibular nerve*.

Incise the radial muscles of one side close to the angle of the mouth. Remove the muscles and clear away the fat deep to display as much as possible of the *buccinator muscle*.

Using the illustration in the atlas, identify as many of the muscles as possible. Pay attention to the relations of the facial artery and vein to the muscles.

34 A

Pull the lower lip downwards and incise the mucosa between the lip and the gum. Find the *mentalis muscle* and note its origin and insertion.

Leave the muscles in the orbital region for the dissection of the orbit and eyelid.

4

The Parotid Region

In this dissection the parotid gland, the parotid duct and their relations are demonstrated. The gland is cleaned, then carefully picked away to demonstrate the structures which pass through it. Finally, the gland is completely removed to demonstrate the structures which form the parotid bed.

Remove the skin of the face as far posteriorly as the auricle.

Find the *zygomatic arch* by palpation . Clean the area between the zygomatic arch and the margin of the mandible as far posteriorly as the auricle to define the *parotid gland*, the *parotid duct* and the *masseter muscle* (**1**).

> *As the gland is cleaned, note that the investing layer of the deep cervical fascia forms a capsule for the parotid gland before it blends with the fascia over the masseter muscle* (**2**).
>
> **34 A** *Define the superficial margins of the parotid gland.*
>
> *Try to find and preserve all the branches of the facial nerve as they leave the anterior margin of the gland.*
>
> *Note where the parotid duct pierces the buccinator muscle.*
>
> **36 A,B** *Note the origin and insertion of the masseter muscle.*

> *Try to find the auriculotemporal nerve and the superficial temporal vessels in front of the auricle.*

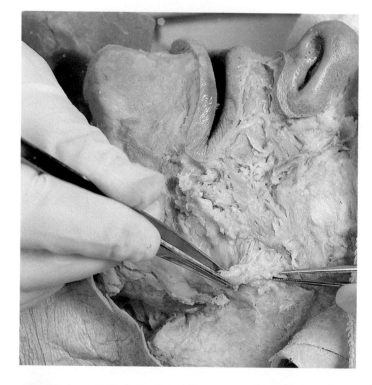

1

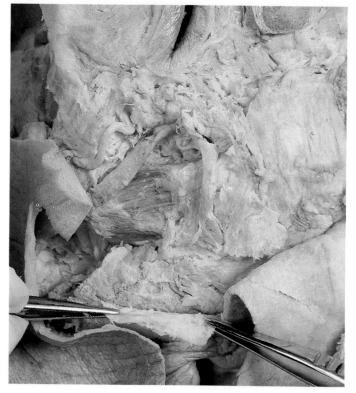

2

16

*Identify the main branches of the **facial nerve**, the **auriculotemporal nerve**, the **external carotid artery** and its terminal branches and the **retromandibular vein**.*

Trace the retromandibular vein down into the neck.

Note the relations of the parotid gland to the masseter and the mandible, and to the sternocleidomastoid muscle.

Start removing the gland piecemeal to demonstrate the structures which pass through it (**3**).

Remove the sternocleidomastoid muscle from its insertion into the mastoid process on one side only (**4**). Pick out the remainder of the parotid gland to display the structures which form the parotid bed.

3

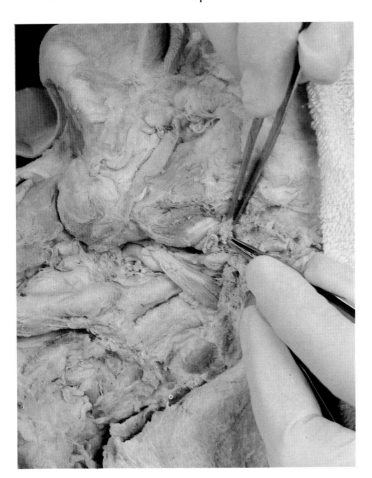

*Find the muscles and ligaments attached to the **styloid process**, the **internal carotid artery** and the **internal jugular vein**.*

It should now be possible to find all the branches of the external carotid artery, except the ascending pharyngeal artery.

*The **superior thyroid, lingual facial** and **occipital arteries** were seen in the dissection of the anterior triangle of the neck. The occipital artery can also be seen at the apex of the posterior triangle.*

*The **posterior auricular artery** may be found running along the superior border of the posterior belly of the digastric muscle.*

*The **maxillary** and **superficial temporal arteries** have been seen in the present dissection.*

4

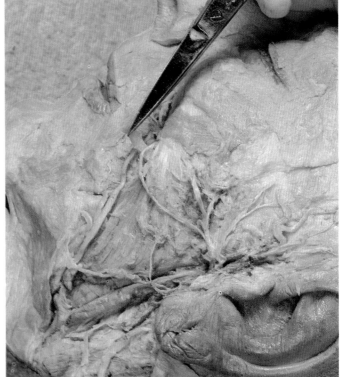

5

The Scalp

A full dissection of the scalp is time consuming and difficult. It is sufficient to observe the layers of the scalp and to construct a diagram of its nerve and blood supply.

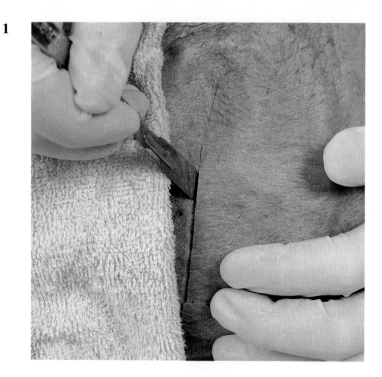

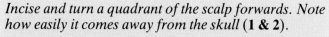

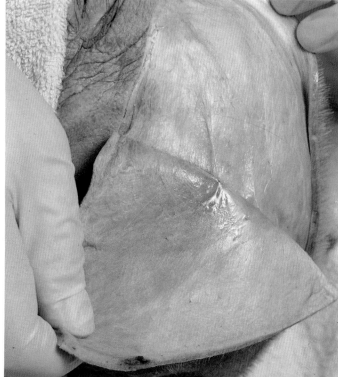

Incise and turn a quadrant of the scalp forwards. Note how easily it comes away from the skull (1 & 2).

Identify the three layers of the scalp in the quadrant you hold in your hand:
- ***skin**, with hair,*
- ***connective tissue**, dense with loculi of fat,*
- ***aponeurosis** and **frontal belly** of the **occipitofrontalis muscle**,*
- ***loose connective tissue**, forming a potential space, the layer at which the three outer layers parted from the skull,*
- ***periosteum**, which still covers the bones of the vault of the skull.*

Explore the potential space between the aponeurosis and periosteum with a blunt probe. Confirm that the space continues into the eyelid.

34 A

Use the illustrations in the atlas and your dissection to name the arteries, veins and nerves which supply the scalp. Trace their courses on the specimen with your finger. Write their names on to the diagram (3).

The part of the superficial dissection of the head and neck is complete. It is a good idea to review all the structures you have demonstrated so far with the aid of the appropriate illustrations in the atlas.

6

The Cranial Cavity

*In this dissection the cranial cavity is opened by removal of the skull cap.
Then the brain is carefully removed, so that its relations to the cranium
and to the membranes which invest it (the meninges) can be studied. The
cranial nerves are followed from their origins through the cranial cavity.
Your attention is directed to the location of the meningeal and cerebral
blood vessels with respect to the meninges.*

It is a good idea to study the osteology of the cranial interior before starting this dissection.

A saw and a mallet and chisel will be required for this dissection.

Cut around the scalp from the *external occipital protuberance* to just above the orbital margin, continue this cut back to the external occipital protuberance. Use this cut as a guide as you saw through the bones of the skull (**1**).

Use the chisel and mallet to free the skull cap, then use the chisel to lever it off (**2 & 3**).

Examine the inner surface of the skull cap and the **dura mater**. *Open the* **superior sagittal sinus** *and note the* **arachnoid granulations**. *Examine the branches of the* **middle meningeal vessels** *on the dura mater and note the grooves on the inner surface of the skull cap which correspond to these vessels.*

54 A

1
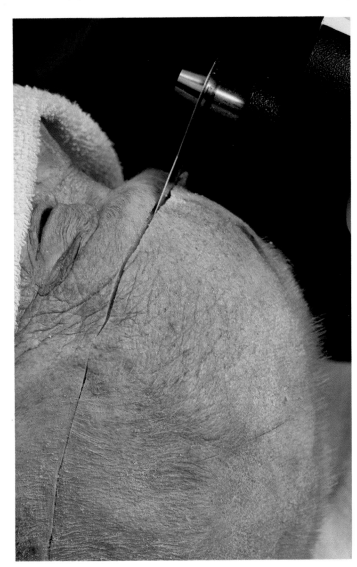

2
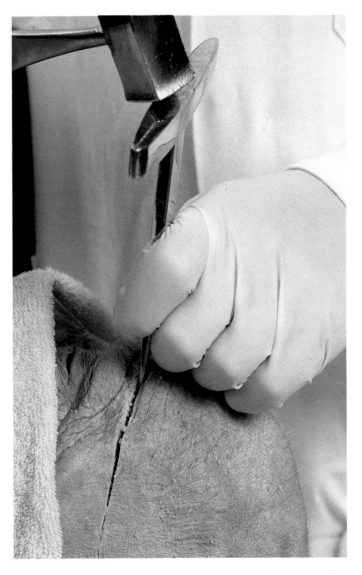

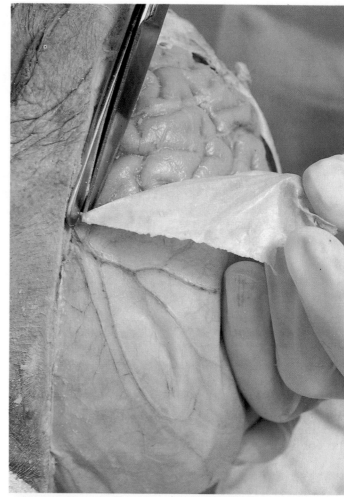

Reflect the dura mater to expose the *cerebral hemispheres* invested by the *arachnoid* and the *pia mater*. Leave a strip of dura mater surrounding the superior sagittal sinus in the midline (**4**).
Use forceps to pick off the arachnoid membrane and the superficial blood vessels of the hemisphere (**5**).

Confirm that the dura mater was firmly attached to the bones of the skull, that there is a potential space between the dura mater and the arachnoid.

Note that there is a definite space between the arachnoid and the brain and that the arachnoid, superficial blood vessels and the delicate pia mater are all removed together as you expose the cerebral cortex of the hemisphere.

Confirm that the meningeal vessels run between the bone and the dura, that the venous sinuses are enclosed within the dura and the superficial vessels of the brain run in the subarachnoid space.

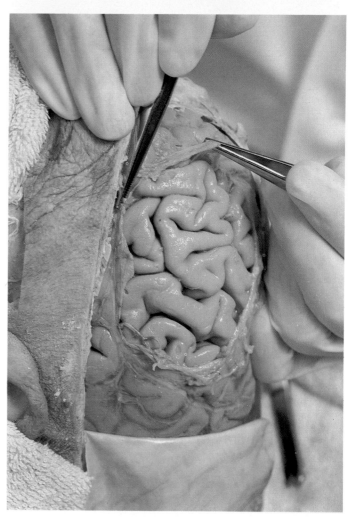

Push one of the hemispheres laterally so that you can examine the *falx* and find the *corpus callosum* which runs between the hemispheres below the falx (**6**).

Cut through the corpus callosum and the brain below it to free the cerebral hemispheres. Lift out the hemispheres one at a time, cutting through the *optic nerves* and *brain stem* as cleanly as possible to release them (**7**).

*Examine the **falx** and **tentorium**, both reflections of the dura mater. Note their attachments to each other and to the walls and floor of the cranial cavity.*

*Identify the **venous sinuses** which run in the falx and tentorium, opening them up if necessary (**8**).*

Cut through the tentorium to expose the *cerebellum* (**9**).

6
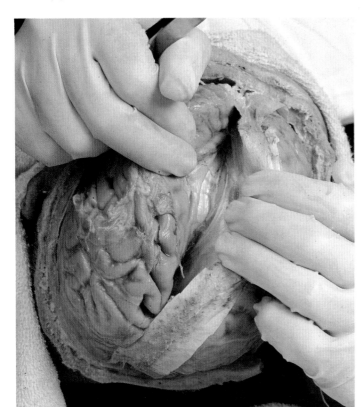

7
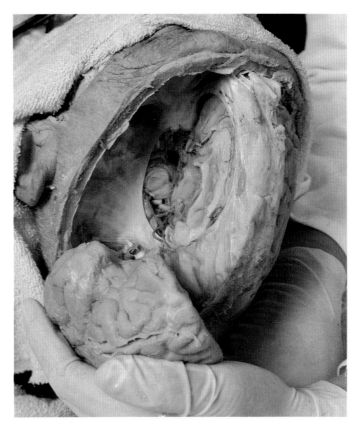

8
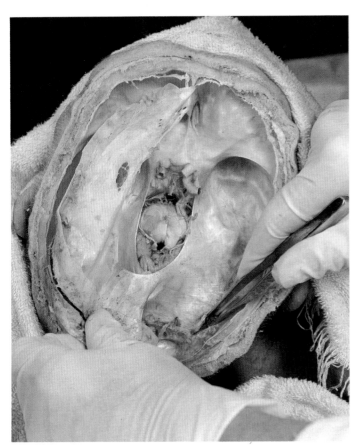

9
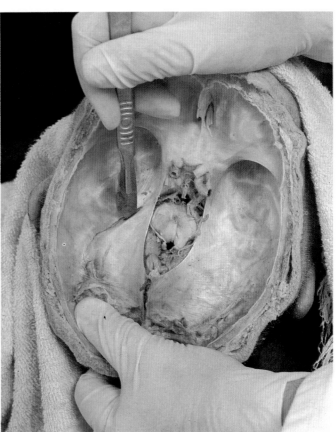

Remove the cerebellum in three or four large pieces, taking care to avoid damage to the brain stem to which it is attached (**10**).

*Very gently, push the brain stem to observe the **third to twelfth cranial nerves**, as they pass from their origins on the surface of the brain stem towards their foramina of exit from the cranial cavity. Identify the three parts of the brain stem, the **midbrain, pons** and **medulla oblongata**, and each of the cranial nerves by name, noting the position of its origin on the brain stem (**11**).*

10

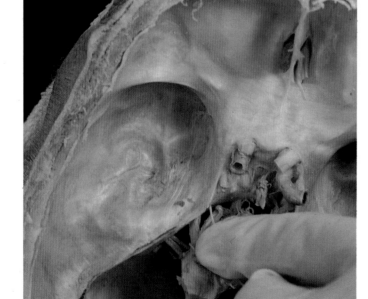

Remove the brain stem by cutting through the cranial nerves as close as possible to it. Finally, cut through the junction between the brain stem and spinal cord at the level of the *foramen magnum.*

*Identify the various parts of the brain, the two cerebral hemispheres and the two halves of the **diencephalon**.*

62 A, B,C

*Find the **frontal, parietal, temporal** and **occipital lobes** of the cerebral hemispheres. Identify again the midbrain pons and medulla oblongata. Reassemble the parts of the brain and indicate the origins of the twelve pairs of cranial nerves.*

63 D, E

67 D

*Now identify the cranial nerves in the cranial cavity as they pass towards their foramina of exit. Note the **spinal root of the accessory nerve** as it ascends from the foramen magnum to join the **cranial root**.*

54 B,C 55 D

*Examine the parts of the **arterial circle** on the floor of the cranial cavity and note the relations of its major tributaries and branches to the cranial nerves.*

Find:
○ *the **two vertebral arteries** and the **basilar artery**, and the roots of:*
○ *the **superior, anterior** and **posterior inferior cerebellar arteries**,*
○ *the **anterior, middle** and **posterior cerebral arteries**,*
○ *the **anterior** and **two posterior communicating arteries**.*

Find the *trigeminal nerve* (**12**). Follow it to the *trigeminal cave* on the floor of the middle cranial fossa. Open the trigeminal cave and find the *trigeminal ganglion*.

54 C

11

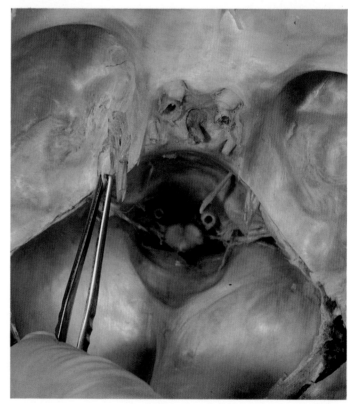

12

54 C *Remove the dura mater from the rest of the floor and wall of the middle cranial fossa on one side only, to find the stems of the middle meningeal vessels and the greater and lesser petrosal nerves (13).*

Remove the dura mater from the medial part of the middle cranial fossa taking great care to avoid damage to the nerves which run within it. In this way you will expose the *cavernous sinus*.

56 A (9,13,14, 15,16) *Define the extent of the cavernous sinus.*

*Find the **oculomotor nerve**, the **trochlear nerve** and the **ophthalmic and maxillary divisions of the trigeminal nerve** as they run through the lateral wall of the sinus.*

*Find the **internal carotid artery** and the **abducent nerve** as they run through the sinus itself.*

55 D (29) Find the remains of the *pituitary stalk* in the midline of the floor of the middle cranial fossa. Make a circular incision in the dura here (*diaphragma sellae*) to expose the *pituitary gland*. Carefully lift out the pituitary gland and try to identify its *anterior* and *posterior lobes*. Make a median sagittal section of it if necessary.

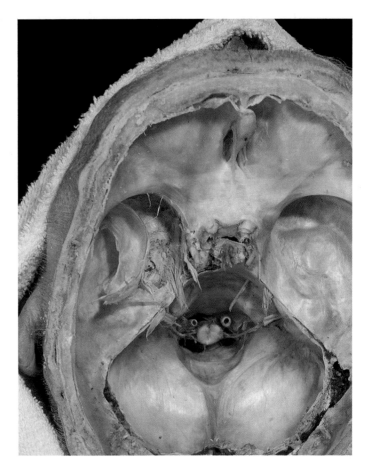

Table of foramina leading from the floor of the cranial cavity		
Foramen	**Destination**	**Contents**
Anterior cranial fossa		
Foramina of the cribriform plate	Nasal Cavity	
Middle cranial fossa		
Optic canal	Orbit	
Superior orbital fissure	Orbit	
Foramen rotundum	Pterygopalatine fossa	
Foramen ovale	Infratemporal fossa	
Foramen spinosum	Infratemporal fossa	
Foramen lacerum		
Hiatus for the greater petrosal nerve	Tympanic cavity	
Hiatus for the lesser petrosal nerve	Tympanic cavity	
Posterior cranial fossa		
Internal auditory meatus	Inner ear and tympanic cavity	
Jugular foramen	Neck (carotid sheath)	
Hypoglossal canal	Neck (carotid sheath)	
Foramen magnum	Vertebral canal	

Use a dry skull to identify the foramina in the posterior and middle cranial fossae. Observe where these foramina lead to, and use the information acquired in this dissection to construct a list of the structures which pass through the foramina (table).

The Orbit and Eyelid

In this dissection the orbit will be opened from above and from the front, in order to demonstrate the extrinsic muscles of the eye and their nerve supply. In addition the eyelids, conjunctival sac, lacrimal gland and the interior of the eyeball will be examined. The ophthalmic division of the trigeminal nerve will be traced into the orbit.

1

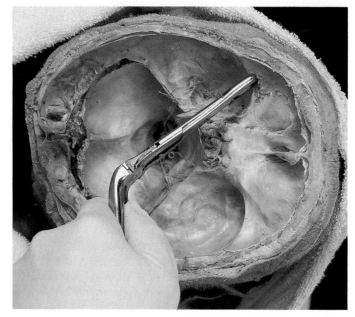

2

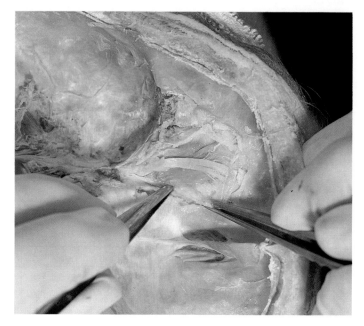

56 A *Before starting to dissect, identify the **third, fourth** and **sixth cranial nerves** and **ophthalmic division of the trigeminal nerve**, as they pass through the cranial cavity and cavernous sinus. In this dissection these nerves will be traced into the orbit.*

Remove the dura mater from the floor of the anterior cranial fossa.

Use a mallet and chisel to crack the bone which makes up the floor of the anterior cranial fossa and the roof of the orbit. The bone is thin, so a gentle tap is usually enough. Use forceps to remove the fragments of bone.

*Identify the **frontal nerve** lying on the orbital fat, just beneath the bony roof of the orbit.*

Use the rongeurs to nibble away the bone at the posterior margin of the anterior cranial fossa (**1**). (This is mostly the lesser wing of the sphenoid bone; check this on a dried skull.) Use the rongeurs to nibble away any sharp pieces of bone which remain on the roof of the orbit.

Use forceps to pick away the fat which encloses the muscles, nerves and vessels of the orbit. Cut *levator palpebrae superioris* and *rectus superior* and reflect them forwards (**2**).

Identify all the orbital structures that can be seen from this aspect. Trace the nerves into the orbit from the cranial cavity; note the arrangement of nerves and vessels in the superior orbital fissure, and where the nerves branch. **56 D,E 57 F**

Now turn your attention to the face of the cadaver.

Fully define the *orbicularis oculi muscle*, and the *frontal belly of occipitofrontalis*. **34 A**

*Explore the **conjunctival sac** with a blunt probe.*

3

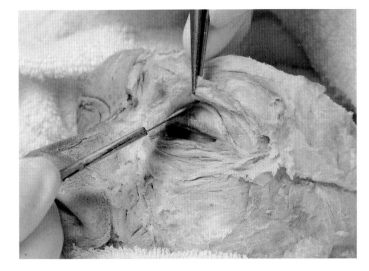

4

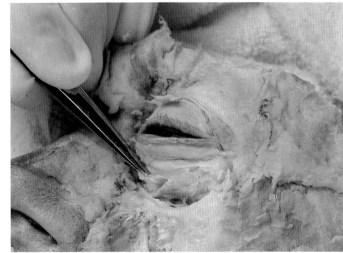

5

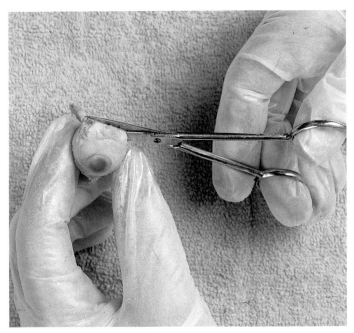
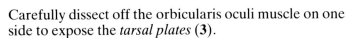

6

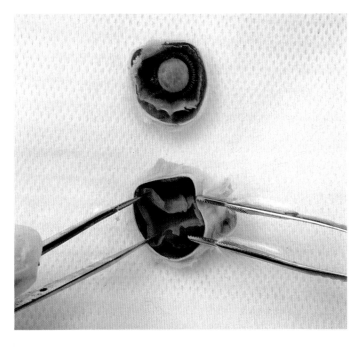

Carefully dissect off the orbicularis oculi muscle on one side to expose the *tarsal plates* (**3**).

*Note the position and relations of the **lacrimal gland**.*

56 C
57 C Break through the *orbital septum* between the inferior tarsal plate and the orbital margin. Remove the orbital fat piecemeal to display the *inferior oblique muscle* and the *inferior rectus muscle* (**4**).

Remove one eyeball by cutting through the muscles as close as possible to their insertions.

56 C *Identify the precise positions of these insertions on the eyeball.*

Make an equatorial cut in the coronal plane, through the eyeball. Wash out the *vitreous humour* gently (**5**).

*Examine the posterior half of the specimen underwater (**6**).*

*Note the layers, the white **sclera**, the dark **choroid** layer and the delicate, easily detached, greyish **retina**.*

Examine the anterior half of the specimen underwater.

*Remove the **lens**, pass a fine probe through the pupil into the **anterior chamber** of the eye. Observe the tip of the probe from the anterior aspect.*

The Ear and Temporal Bone

The purpose of this dissection is to display as much as possible of the anatomy of the external, middle and inner ear. Most of the dissection is concerned with the contents and walls of the tympanic cavity.

This dissection is possible if an isolated, decalcified, temporal bone is available.

58 A

*Examine the external ear and name its parts. Orientate the specimen carefully, and identify any nerves and vessels still attached to it. (These may include the **facial**, intermediate and **vestibulo-cochlear** nerves, the **greater and lesser petrosal nerves**, the **internal carotid artery**.)*

Remove the auricle and the squamous part of the temporal bone. Clear the wax from the external auditory meatus. Remove the *tegmen tympani* to open the *tympanic cavity* (**1**). Continue the removal of bone posteriorly to open the *mastoid antrum*.

59 E,F, G,H

*Inspect the **tympanic cavity** and **mastoid antrum**. Identify the three **ossicles**.*

Remove the *incus* and *stapes*, using a mounted needle and fine forceps. Gradually pare down the roof of the *internal auditory meatus* and find the *vestibulocochlear* and *facial nerves*.

Identify the *auditory tube* (**2**). Divide the bone along the line of the auditory tube.

58 C,D

Inspect the medial half of the specimen and identify the auditory tube, the structures on the medial wall of the tympanic cavity, the mastoid antrum and air cells.

*Follow the **facial nerve** through the **facial canal** towards the **stylomastoid foramen**.*

*Inspect the lateral half of the specimen. Identify again the **malleus** and the **tympanic membrane**.*

Cut the cartilaginous part of the external auditory meatus away. Note the direction taken by the external auditory meatus.

Hold the specimen up to the light and observe the tympanic membrane through the external auditory meatus.

58 B

*The **arcuate eminence** is above the **anterior semicircular canal**. The **tympanic promontory** shows the position of the cochlea.*

Using the above information slice the bone carefully, to see the canals of the *bony labyrinth*. Look for the *membranous labyrinth* inside these canals (**3**).

59 E,F G,H

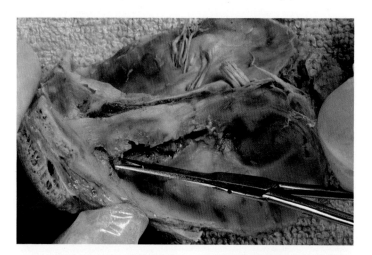

1

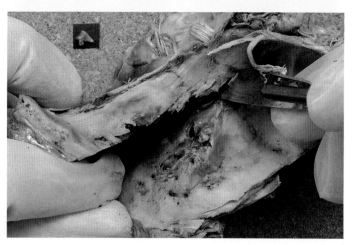

2

3

The Temporal and Infratemporal Fossae

This dissection requires the reflection of the temporal and masseter muscles as well as the removal of some bone from the zygomatic arch and the ramus of the mandible to display the medial and lateral pterygoid muscles. The main branches of the mandibular division of the trigeminal nerve and the first and second parts of the maxillary artery should then be followed through the infratemporal fossa.

1

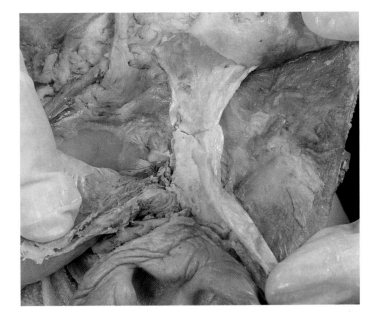

2

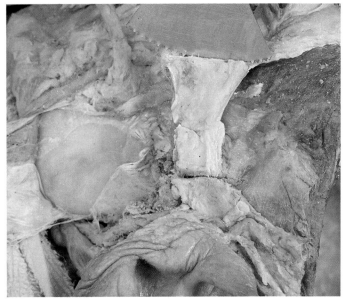

13 C

The infratemporal fossa is situated between the ramus of the mandible and the base of the skull. It is a good idea to study the walls and roof of the fossa in a dried skull.

The infratemporal fossa does not have a bony floor or posterior wall.

Identify all the foramina and fissures through which the other craniofacial regions can communicate with the infratemporal fossa.

Keep the dry skull on the dissecting table, as you will need to refer to it from time to time during the dissection.

Find the *zygomatic arch* on one side of the cadaver by palpation.

Reflect and then dissect away the *temporalis fascia* above the zygomatic arch to display the *temporalis muscle* (**1**).

Note the directions taken by the fibres in the anterior and posterior parts of the muscle.

Since only the lower part of the muscle is present on dissection, trace the full extent of the temporal fossa and origin of the muscle on the dry skull.

Detach the *masseter muscle* from its origin on the zygomatic arch and turn it downwards, stripping it away from the ramus of the mandible.

Divide the zygomatic arch just posterior to the body of the *zygoma* and just anterior to the *temporomandibular joint* (**2**). Remove the piece of bone.

Observe the insertion of the temporalis muscle into the ***coronoid process*** *of the mandible.*   **43 B**

Divide the coronoid process of the mandible and turn the temporalis muscle upwards.

3

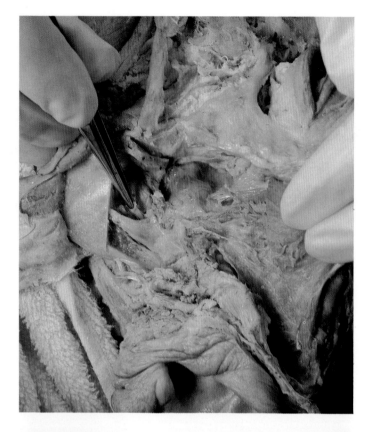

Divide the ramus of the mandible:
- through the *condylar process* just below the temporomandibular joint, and
- along a line continued posteriorly from the upper border of the body of the mandible. The purpose of the placement of this second cut is to expose as much as possible of the contents of the infratemporal fossa while avoiding damage to the structures which pass towards the *mandibular foramen* on the medial side of the ramus of the mandible.

Remove the bone fragment. The periosteum should remain in place protecting the contents of the fossa. Pick off the periosteum carefully.

*Inspect the infratemporal fossa before proceeding with the dissection. Identify the **inferior alveolar nerve and vessels** and the **sphenomandibular ligament** (3).*

Demonstrate the attachments of the sphenomandibular ligament and the course of the inferior alveolar nerve on the dried skull.

4

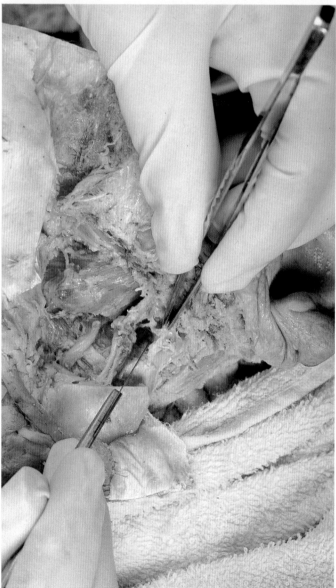

Using the illustration in the atlas as a guide, carefully pick away the veins, fat and other loose tissue to display the *medial* and *lateral pterygoid muscles*. Clean the *maxillary artery* and the *branches of the mandibular nerve* in the fossa and follow the *buccal nerve* into the cheek (4).

37 A
B,C

*Study the origins and insertions of the **pterygoid muscles** in the dissection and on the dry skull.*

Identify the nerves and study their courses noting their relations to the muscles.

Identify the first part of the maxillary artery, lying deep to the neck of the mandible. Find as many of its branches as possible.

*Follow the second part of the maxillary artery through the infratemporal fossa towards the **pterygopalatine fossa**.*

Find the pterygopalatine fossa on the dry skull.

16 A
24 A
B,C

The deeper structures in the fossa and the temporomandibular joint will be studied in the dissections of the pharynx.

Repeat the dissection on the opposite side of the head.

10

Dissection of the Pharynx from the posterior aspect

The pharynx lies posterior to the nasal and oral cavities and the larynx. Probably the best way to appreciate the shape and relations of the pharynx is to separate it from the posterior structures of the head and neck and to view it from its posterior aspect.

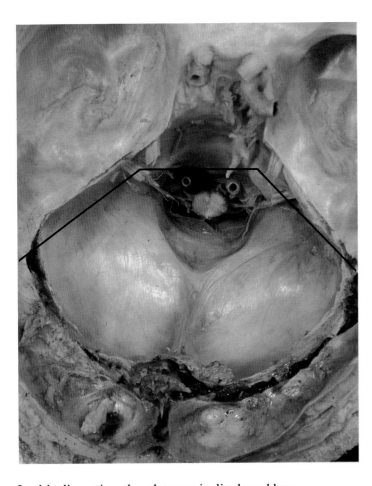

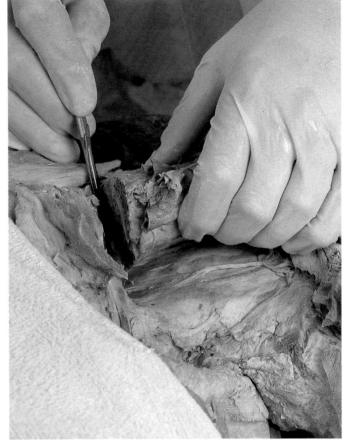

In this dissection the pharynx is displayed by:
- sawing through the base of the skull,
- dividing the visceral structures of the neck at the superior aperture of the thorax,
- separating the pharynx from the posterior parts by blunt dissection,
- dividing the muscles and ligaments which attach the more anterior parts of the skull to the vertebral column.

Finally, the pharynx is opened so that its anterior wall can be studied.

Make a saw cut through the *basilar part of the occipital bone* just anterior to the *foramen magnum*. Continue the cut laterally on both sides, avoiding the hard *petrous part of the temporal bone* (**1**).

Cut through *the trachea, oesophagus* and both *carotid sheaths* as close as possible to the *manubrium of the sternum* (**2**).

Use your hands to separate the oesophagus and the pharynx from the *prevertebral fascia* between the superior aperture of the thorax and the base of the skull. Identify the *cervical sympathetic trunk* and its *ganglia*, and leave them on the prevertebral fascia.

3

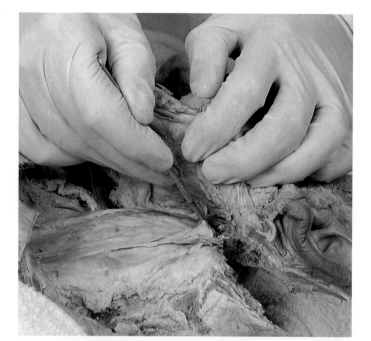

The two parts of the dissection are now connected only by the muscles and ligaments which pass between the part of the skull anterior to the saw cut and the cervical vertebral column. Cut through these muscles and ligaments (**3 & 4**).

Clean the *posterior surface of the pharynx* (**5**).

Use the illustration in the atlas to identify the **pharyngobasilar fascia** *and the* **constrictor muscles** *of the pharynx.*   **50 A**

Use forceps and blunt dissection to open the carotid sheaths and to display as many of the nerves and vessels as possible.

Use the illustration in the atlas to identify as many of the lateral relations of the pharynx as is possible.   **50 A**

4

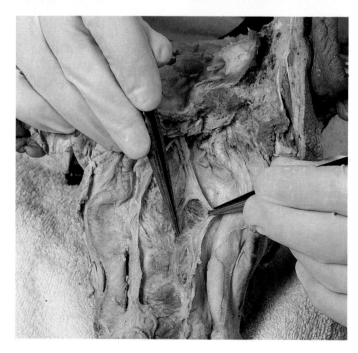

Use scissors to make a midline incision through the full length of the posterior wall of the pharynx (**6**). Wash out the pharynx with running water.

Inspect the anterior wall of the pharynx without disturbing any of its parts.   **51 B**

Note those structures which can be seen in the **nasopharynx**, *the* **oropharynx** *and the* **laryngopharynx**.

Inspect the interior of the larynx and identify the **vestibular** *and* **vocal folds**. *Find the* **epiglottic valleculae** *and the* **piriform fossae**.   **52 J,K**

If your specimen allows an easy approach to the soft palate, dissect away some of the mucous membrane on its superior surface to try to demonstrate the *levator veli palatini* and trace this muscle laterally.   **51 B**

5

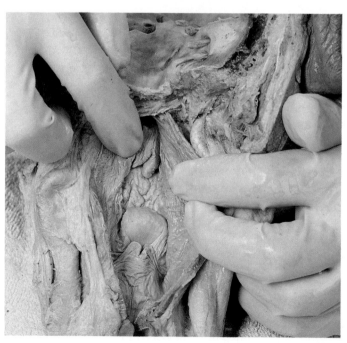

6

11

The Deep Neck, or Viscera of the Neck

The purpose of this dissection is to display the thyroid gland and its vessels, and to look at the larynx from its anterior aspect.

38 A *Identify the muscles on the **anterior aspect** of the specimen which were studied during the dissection of the anterior triangle of the neck.*

Reflect the omohyoid, sternohyoid and sternothyroid muscles upwards. Clean the fat and loose tissue from the front of the trachea, preserving any vessels that you find.

Remove the *pretrachial fascia* from the front of the *thyroid gland* (**1**). Clean all the blood vessels associated with the thyroid gland and trace the lateral ones back to their trunks. The *inferior thyroid arteries* and *veins* have been divided close to the superior aperture of the thorax.

39 B *Identify the **lobes** and **isthmus** of the **thyroid gland** and all its vessels.*

*Pull the inferior part of one lateral lobe of the thyroid away from the **trachea** and **oesophagus**, and try to find the **recurrent laryngeal nerve** running in or close to the groove between the trachea and the oesophagus.*

Divide the isthmus of the thyroid and the lateral vessels on one side and remove one lobe of the thyroid gland.

*Examine the posterior surface of the lobe and try to identify the **parathyroid glands**.*

44 A Clean the front of the larynx and trachea. Clean the recurrent laryngeal nerves.

39 B
52 A *Identify all the cartilages of the larynx and the **cricothyroid muscles**.*

53 A *Find the **arytenoid** and **epiglottic cartilages** on the posterior aspect of the larynx by palpation, and correlate your findings with the photographs of the skeleton of the larynx.*

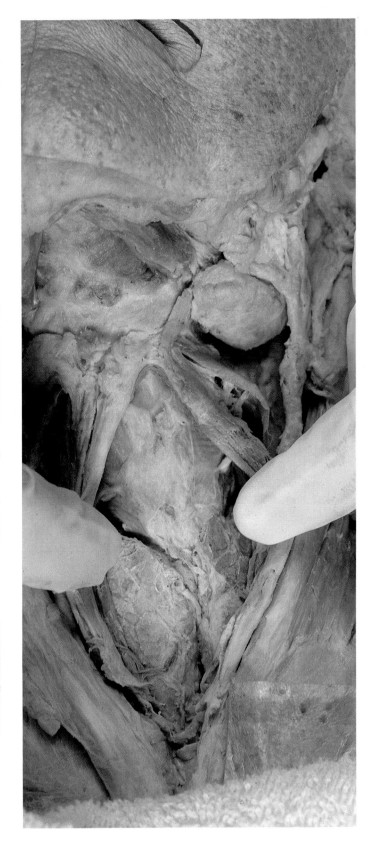

1

12

Dissection of the Larynx and Pharynx from the lateral aspect

In order to display the lateral aspect of the larynx and pharynx, it is necessary to remove almost half of the mandible and the muscles of the infratemporal fossa. This dissection also displays the submandibular region. The lingual and hypoglossal nerves can be traced forwards towards the mouth.

Carry out this dissection on one side of the specimen only, retaining the best dissection of the infratemporal fossa.

Dissection of the temporomandibular joint
Clean the posterior end of the *lateral pterygoid muscle* and try to define its insertions. Clean the *capsule* of the joint.

> *Observe the **intra-articular disc** and the two cavities of the joint.*

Disarticulate the temporomandibular joint by cutting all the way around the capsule and its associated ligaments. Divide the lateral pterygoid muscle where it inserts into the joint.

> *Inspect the articular surfaces and the intra-articular disc.*

Removal of the mandible
Pull the mandible towards you and cut through the following structures as close to the mandible as possible (**1**):
- the *buccinator muscle* and the muscles of the lower lip,
- the *medial pterygoid muscle* and the *stylomandibular ligament*,
- the *mylohyoid muscle* and the mucous membrane of the mouth.

Divide the mandible about half an inch lateral to the midline. Pull the mandible gently away from the specimen. It may be necessary to cut through the mandibular origins of the *digastric geniohyoid* and *genioglossus muscles*, if your section through the mandible is too close to the midline.

43 B
44 A
> *Study the floor of the mouth. The **mylohyoid nerve** and **vessels** may be attached to the inner surface of the mandible.*

Remove the lateral and medial pterygoid muscles from the *lateral pterygoid plate* and adjacent bones. Be careful to preserve all the arteries and nerves which pass between them. Remove the *infrahyoid strap muscles*, except the *thyrohyoid muscle*.

Trace the *superior laryngal nerve* and its branches to the larynx. Trace the *hypoglossal nerve* into the floor of the mouth. Follow the *lingual nerve* from the infratemporal fossa into the floor of the mouth. Use forceps and blunt dissection to display as many of the branches of other nerves and vessels as you can. Clean the sides of the pharynx and larynx. Display the *tensor veli palatini* muscle.

37 C
> *Study the specimen with the aid of the atlas.*
>
> *Look for the **otic ganglion** deep to the mandibular nerve. Observe the **auriculotemporal nerve**, **middle meningeal artery** and the **chorda tympani nerve**, and their relations to the **spine of the sphenoid bone**.*
>
> *Delineate the attachments of all muscles encountered in this dissection on a dry skull.*

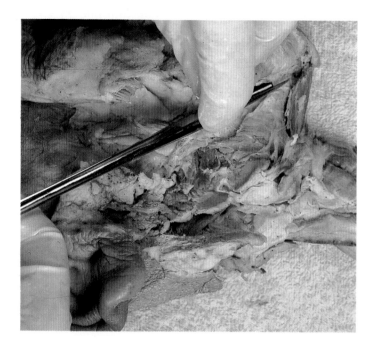

1

13

The Prevertebral Region

The purpose of this dissection is to display the prevertebral and paravertebral muscles which lie deep to the prevertebral fascia, and to find the cervical sympathetic trunk and ganglia. The dissection also displays the muscles which form the floor of the posterior triangle of the neck and the upper part of the brachial plexus. Particular attention is given to the scalenous anterior muscle and its relations which are important because this muscle lies at the junction between the neck, upper limb and thorax.

1

2

44 B *Look at the **prevertebral muscles** through the prevertebral fascia. The **longus colli** is the medial muscle. **Longus capitis** is more lateral and inserts into the skull. Do not worry about the precise attachments of these muscles.*

*Find the **scalenus anterior** and **brachial plexus** deep to the **prevertebral fascia**. Observe that the prevertebral fascia is continued laterally over the brachial plexus as the **axillary sheath**.*

44 B *Find the **cervical sympathetic trunk**; if it has become detached replace it in its proper position and identify the **superior** and **middle cervical ganglia**. Note the relations of the cervical sympathetic trunk, and the size and vertebral levels of the ganglia.*

It is important to preserve the relations of the *scalenus anterior*, so use the illustration in the atlas as a guide while you dissect (**1**). Remove the prevertebral fascia with great care (**2**). Develop the planes between the scalene muscles and brachial plexus using your fingers only.

44 B *Study the relations of the **scalenus anterior muscle**.*

Move the scalenus anterior muscle medially, still taking care to preserve its relations. Separate the contents of the axillary sheath using blunt dissection with closed forceps as far as possible.

*Identify the **roots** and **trunks of the brachial plexus**, noting their positions in the neck. Identify the **subclavian artery** and **vein**. Find the insertion of the scalenus anterior muscle into the first rib.*

Use your fingers to develop the planes between the *levator scapulae* and surrounding muscles.

*Leave the **trapezius muscle** as it is and revise the position of the sternocleidomastoid muscle. Identify the **scalene muscles**, **levator scapulae** and **splenius capitis**, confirming that these muscles form the floor of the posterior triangle of the neck.* **43 B**

*Replace the facial part of your dissection. If the **accessory nerve** remains on the stump of insertion of the sternocleidomastoid to the mastoid process, reconstruct the path of the accessory nerve from the carotid sheath, deep to the sternocleidomastoid, across the roof of the posterior triangle and into the trapezius muscle.*

The Root of the Neck

The purpose of this dissection is to study the arrangement of structures found at the junction between the neck and thorax. It is also a good time to bring together the information gathered in the earlier dissections of the neck.

1

Look at an assembled skeleton and observe the bones which surround the superior aperture of the thorax.

Revise the attachments and relations of the scalenus anterior in the dissection. Trace the phrenic nerve into the thorax.

Clean the structures at the superior aperture of the thorax, by blunt dissection and by removing loose connective tissue with forceps (1).

151 F

Identify the structures at the root of the neck, noting the relation of the subclavian vessels to the **suprapleural membrane.** *Deep to this membrane is the* **apex of the lung.**

Find as many of the branches of the **subclavian artery** *as you can.*

Use the dissection to identify the neural and vascular structures related to the **first rib.**

Finally, use the information you have gathered in this and in earlier dissections to make a diagram of a transverse section through the neck, not forgetting the layers of cervical fascia.

2

Match the structures found in the deep dissection of the neck with the structures at the root of the neck (2).

15

Dissection of the Pharynx, Nasal Cavity and Oral Cavity from the medial aspect

In this dissection the specimen will be bisected in the median sagittal plane, in order to study the pharyngeal, nasal, oral and laryngeal cavities from the medial aspect. The pterygopalatine fossa will be displayed from the medial aspect. Finally, some of the structures encountered in this dissection will be traced into the craniofacial regions studied in earlier dissections.

Bisect the specimen in the median sagittal plane using a sharp scalpel for the larynx and other soft structures, and a saw for the bone (**1**). Wash out the medial aspect of the dissection under running water.

1

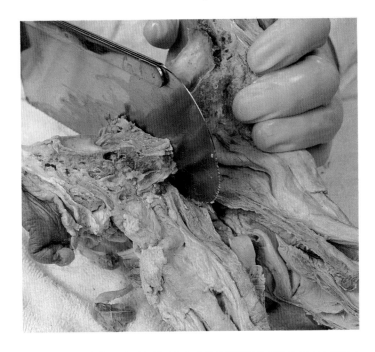

Spend some time studying the medial surface of the specimen without disturbing its parts.

*Firstly, identify the **pharynx** and its subdivisions: the* **53 D** *nasopharynx, oropharynx** and **laryngopharynx**. Note the relation of the subdivisions to the **nasal cavity**, **oral cavity** and the **larynx**.*

Use the illustrations in the atlas to help you find the following structures:
47 C *Lateral wall of the nasal cavity: **superior, middle** and **inferior conchae** and **meati**. The **sphenoethmoidal** recess. Find also the **sphenoidal sinus** in the **body of the sphenoid bone**.*

*Nasopharynx: the opening of the **auditory tube**. The* **47 D** *lateral recess** of the pharynx. The **soft palate** and **uvula**.*

*Oral cavity: the **hard palate**, bone and mucoperiosteum. The **vestibule** of the mouth. The* **46 A** *tongue, the mobile anterior two-thirds. **Genioglossus** and **geniohyoid muscles**. The **mylohyoid muscle**.*

*Oropharynx: the **palatoglossal** and **palatopharyngeal*** **46 A** *folds.*

*Study the following structures: the posterior third of the tongue and the **epiglottic valleculae**.*

*Larynx: the **epiglottis, inlet** and **vestibule** of the larynx, the **vestibular** and **vocal folds**. The **ventricle**, the **infraglottic compartment**, the **lamina** of the **cricoid cartilage**, the **trachea** and **oesophagus**.*

The larynx and pharynx have now been studied from all aspects. Do not try to dissect the intrinsic muscles of the larynx; these are best studied using the illustrations in the atlas when the three-dimensional structure of its cartilages is appreciated.

Dissection of the lateral wall of the nasal cavity
Strip the mucous membrane away from the nasal septum.

*Identify the **perpendicular plate of the ethmoid bone**,* **47 B** *the **vomer** and the **septal cartilage**.*

Crack the septum and remove it piecemeal from the mucous membrane of the opposite side, taking care to avoid damage to the *nasopalatine nerve* and the *sphenopalatine artery*.
Preserve the nasopalatine nerve and sphenopalatine artery, but remove the rest of the mucous membrane to display the lateral wall of the nasal cavity.

Identify the structures shown in the illustrations in the atlas. Note that the vestibule of the nose is lined by **47 C** *skin with stout hairs and that the rest of the nasal cavity and nasopharynx are lined with respiratory epithelium.*

Note that the pharyngeal orifice of the auditory tube is in line with the nostril and inferior meatus.

Confirm that the sphenoidal sinus opens into the sphenoethmoidal recess.

Cut off the inferior concha (**2**).

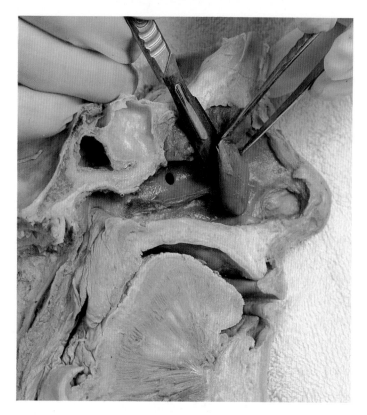

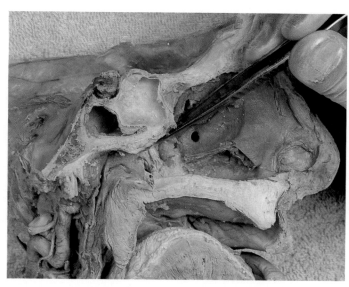

3

48 A *Find the opening of the **nasolacrimal duct**, and pass a fine probe along the duct to demonstrate its continuity with the **lacrimal sac** in the orbit.*

Break away the thin bone over the nasolacrimal duct.

Cut away the middle and superior conchae.

48 A *Observe the openings of the **maxillary sinus** and **hiatus semilunaris** in the middle meatus.*

Dissection of the pterygopalatine fossa
Trace the nasopalatine nerve and sphenopalatine artery back to the lateral wall of the nasal cavity. In this way you will locate the *sphenopalatine foramen*. Strip the mucoperiosteum away from the bone below the sphenopalatine foramen.

*The **greater palatine nerve** can be seen shining through a thin plate of bone; this is the **vertical plate of the palatine bone**.*

Break through this plate of bone using the points of closed scissors. Pick off the fragments of bone to display the *greater palatine nerve* and *artery* in the *greater palatine canal* (**3**).

*Superiorly in the canal, where the nasopalatine nerve joins the greater palatine nerve, there is the **pterygopalatine ganglion**. The other branches of the pterygopalatine ganglion are too fine to be demonstrated by dissection. These nerves and their companion vessels correspond to the canals described in the osteology section.*

At this point it is a good idea to demonstrate the position of the pterygopalatine fossa on a dried skull, to reinforce your understanding of its proximity to the infratemporal fossa, nasal cavity and tonsillar fossa.

Dissection of the floor of the mouth
Pull one half of the tongue towards you (**4**). Carefully remove the mucous membrane from its inferior surface and the floor of the mouth.

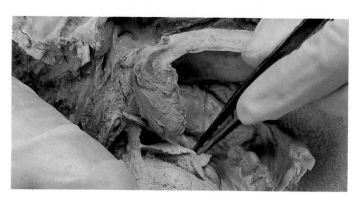

4

*Observe the **mylohyoid muscle** from its superior aspect. Find the **deep parts of the submandibular and sublingual salivary glands**. Locate the **hypoglossal** and **lingual nerves**.*
*Find the **hyoglossus muscle** and trace the **lingual artery** as it passes deep to the hyoglossus muscle.*
*Trace the deep part of the **submandibular salivary gland** around the free edge of the mylohyoid muscle to the **superficial part** in the submandibular triangle.*
*Trace the **lingual nerve** back to the infratemporal fossa, noting its relations in the mouth. Then trace the **mandibular nerve** through the foramen ovale and into the cranial cavity.*

45 C (Lower part)

16
Osteology of the Neck

First, the bones will be studied individually, and then in relation to the articulated skeleton. Finally, some of the soft structures noted in the dissection will be related to the bones in the articulated skeleton.

The individual bones:
- On the skull, find the **mastoid process** of the temporal bone and the parts of the **occipital bone**, especially the **nuchal lines**. Identify the body and ramus of the mandible.

- In the pectoral girdle find the clavicle and the main features of the manubrium of the sternum and the upper part of the scapula.

- The third, fourth, fifth and sixth cervical vertebrae are quite typical. Study one of these, as well as the details of the atlas, axis and seventh cervical vertebra.

- In addition, study the hyoid bone and the first rib.

Use an assembled skeleton to study the way that these bones articulate with each other.

In the articulated cervical vertebral column, note the **intervertebral discs**, the curvature of the column and the **intervertebral foramina** through which the nerves pass from the vertebral canal into the neck.

Pay particular attention to the parts of the vertebra and intervertebral disc which make up the margin of an intervertebral foramen, or are closely related to it.

Note the bones contributing to the **superior aperture of the thorax**. Structures pass between the thorax and the neck or upper limb through the medial part of this aperture. Laterally, the aperture is closed off by the suprapleural membrane **(1)**.

Note the bones which contribute to the **cervicoaxillary canal**, the first rib, the clavicle and the scapula. Structures pass through this canal to the upper limb, from the neck or thorax **(2)**.

Reconstruct the positions of some of the soft parts:
- Trace the position of the **sternocleidomastoid muscle**, between the superior nuchal line and the mastoid process, to the medial part of the clavicle and the manubrium of the sternum.

- Trace the position of the **carotid sheath** from the superior aperture of the thorax, just behind the sternoclavicular joint, to the jugular foramen and carotid canal on the base of the skull.

- Trace the course of the **brachial plexus** from the interverebral foramina between the fourth cervical and first thoracic vertebra across the first rib and into the upper limb through the cervicoaxillary canal.

- Trace the position of the **scalenus anterior muscle**, between the anterior tubercles of the transverse processes of the typical cervical vertebrae, to the scalene tubercle on the first rib.

1

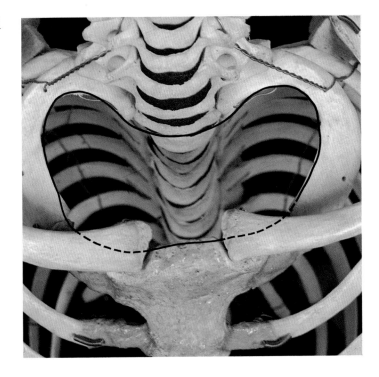

2

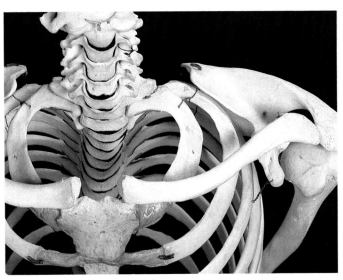

Surface Anatomy of the Neck

In the first instance, observe the general contours of the neck:
○ the **sternocleidomastoid muscles,**
○ the **trapezius muscles** and the **clavicles** forming the borders of the posterior triangle of the neck. The sternocleidomastoid muscle can be made to stand out by asking the subject to look upwards and laterally.

Sometimes the **inferior belly of the omohyoid muscle** can be seen as it crosses the posterior triangle.

Now stand behind the seated subject and palpate the following bony points, using the tips of several fingers **(1)**:
○ the **superior nuchal line** and the **external occipital protuberance,**
○ the **mastoid process of the temporal bone,**
○ the **posterior border of the ramus of the mandible** and the **inferior margin of the body of the mandible,**
○ the **hyoid bone,** palpated between your index finger and thumb.

The **spine of the seventh cervical vertebra** can be felt, but it is usually the **spine of the first thoracic vertebra** which can be seen.

The **lateral mass of the atlas** may be felt by pressing the neck just below the mastoid process.

The pulses of the **common, internal and external carotid arteries** can be felt by pressing the tips of your fingers into the carotid triangle, that is, in the angle between the sternocleidomastoid muscle and the inferior margin of the mandible.

The **facial artery** can be felt as it crosses the inferior margin of the mandible, just anterior to the insertion of the masseter muscle.

Locate the **subclavian artery** by pressing down behind the clavicle.

The jugular pulse **(internal jugular vein)** can be seen close to point where the carotid pulse was felt, and the external jugular vein can be seen when the subject lies down.

You should know where to palpate the main groups of lymph nodes. These are not palpable unless they are swollen as a result of their involvement in disease. Use the illustration to indicate on your subject the location of:
○ the **occipital nodes (a),**
○ the **submandibular nodes (b),**
○ the **submental nodes (b),**
○ the **jugulodigastric (c)** and **jugulo-omohyoid (d) nodes.**

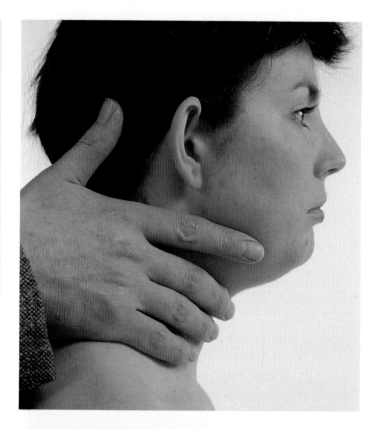

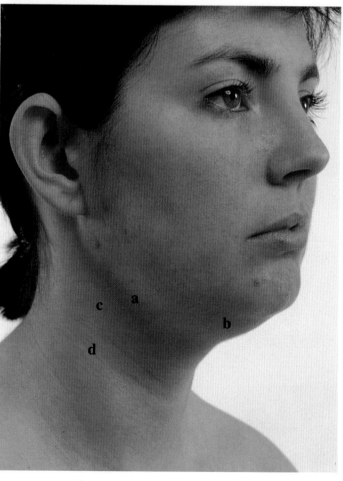

Ask the subject to extend his or her neck, and palpate the following structures in the midline from the point of the chin to the suprasternal notch (3):

a *the symphysis menti,*
b *the myohyoid muscle,*
c *the hyoid bone,*
d *the thyrohyoid membrane,*
e *the thyroid cartilage (feel its notch, prominence and laminae),*
f *the cricoid cartilage,*
g *the trachea,*
h *the isthmus of the thyroid gland,*
i *the trachea,*
j *the suprasternal notch, between the medial ends of the clavicles and the top of the sternum.*

Take note of the position of the arch of the cricoid cartilage. This is at the level of the body of the sixth cervical vertebra.

At this point the larynx leads into the trachea, the pharynx into the oesophagus, the intermediate tendon of the omohyoid muscle crosses the carotid sheath and the inferior thyroid artery passes behind it.

The common carotid artery may be compressed against the transverse process of the sixth cervical vertebra here. This is also the level at which the middle cervical ganglion is found.

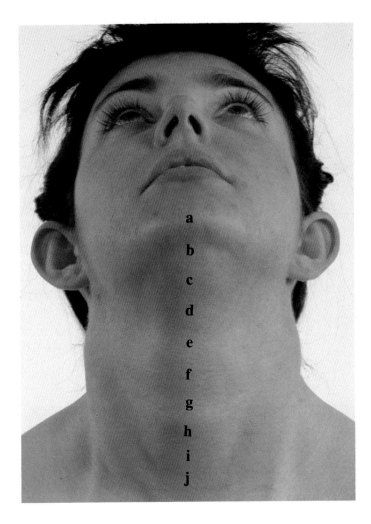

18

The Skull

The purpose of this set of instructions is to provide an overview of the skull which will make further detailed study more profitable.

The first stage is to divide the skull into craniofacial regions: some of these are obvious, for example, the orbits and the nasal cavity. The pterygopalatine fossa and the pharyngeal region are examples of less familiar regions.

We shall then see how the different regions relate to each other and communicate through the foramina, canals and fissures. Finally, we shall study three complicated bones which contribute to several craniofacial regions; these will have to be examined from various aspects.

1

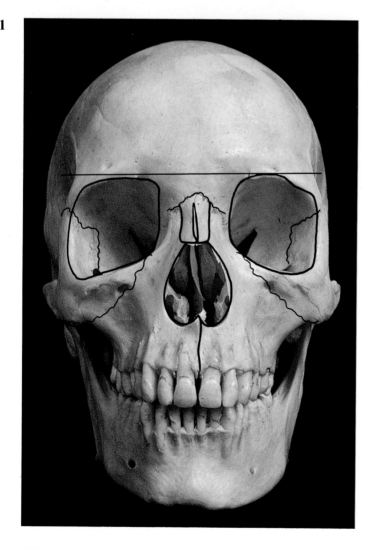

2

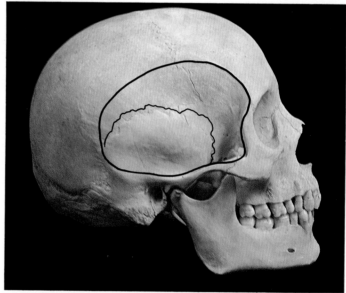

3

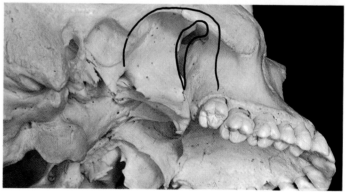

*The skull, seen from the front: find the **vault**, the **orbits**, the **nasal cavity** and the **facial** skeleton. Define the walls and margins of the orbit and the nasal cavity; find the **nasal**, **zygomatic** and **maxillary** bones of the face (**1**).*

*The skull, lateral view: find the vault, **temporal fossa**, **external auditory meatus** and the various parts of the temporal bone (**2**).*

*Define the walls of the **infratemporal and pterygopalatine fossae** (**3**).*

The inferior surface of the skull has a rather complex appearance. However, anteriorly, the **palate** related to the oral cavity and the **posterior apertures of the nasal cavity** are obvious.

Posteriorly, a large area is related to the musculoskeletal compartment of the neck. This is emphasized by the articulation of the upper cervical vertebrae of the important neurovascular bundle, the **carotid sheath** (4).

Less familiar is the **pharyngeal region** which should be traced laterally from the **pharyngeal tubercle** and forwards to the **medial pterygoid plate**. The **infratemporal fossa** extends laterally from the pharyngeal region to the infratemporal crest.

4

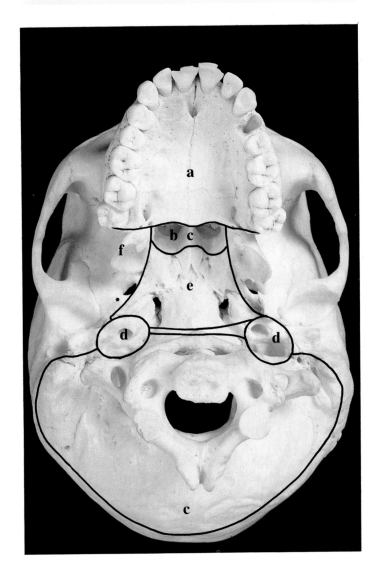

a. Palate
b. Posterior apertures of the nose
c. Musculoskeletal compartment
d. Carotid sheath
e. Pharyngeal region
f. Infratemporal fossa

Define the borders of the three cranial fossae (**5**).

5

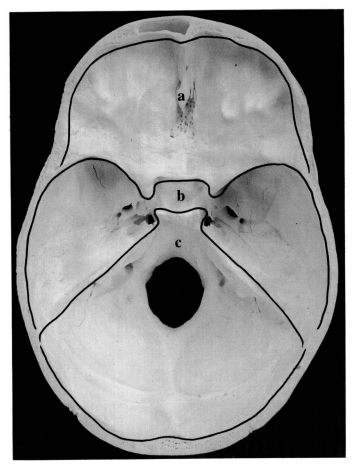

a. Anterior cranial fossa
b. Middle cranial fossa
c. Posterior cranial fossa

The next stage is to relate each of the regions to each other.
Start with the orbit: note that above the roof of the orbit there is the anterior cranial fossa, and below the floor there is the maxillary sinus. The temporal fossa is located laterally to the lateral wall and the ethmoid air cells lie medially to the medial wall.

In a similar way, note the regions related to the nasal cavity, the oral cavity, the pharynx and to each of the fossae of the cranial cavity.

Having defined the main craniofacial regions and noted their relations to each other, use a bristle to study the communications between the regions.

Start with one region, for example, the orbit. Use the atlas to identify the individual foramina, canals or fissures in the region. The superior orbital notch, for instance, runs from the orbit to the upper part of the face, the inferior orbital canal from the orbit to the lower part of the face.

The **nasolacrimal canal** runs from the orbit to the nasal cavity; it enters the cavity below the **inferior concha** (6 & 7).

This is how to find the **pterygoid canal**, at the base of the pterygoid process (8).

6

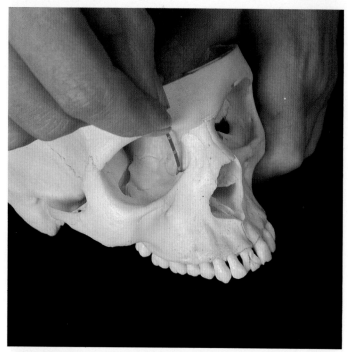

8

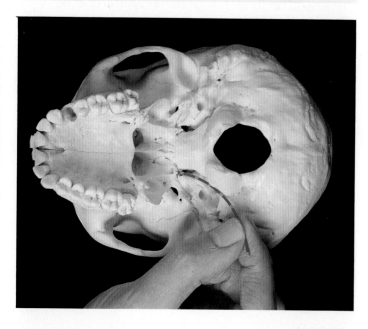

This is how to find the **auditory tube**. Its cartilaginous part is highlighted in blue (9).

7

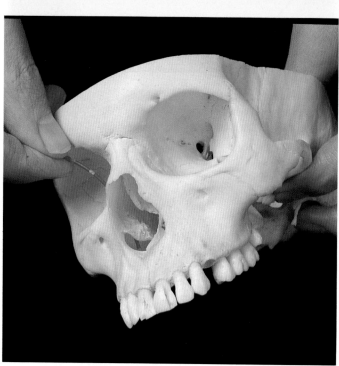

9

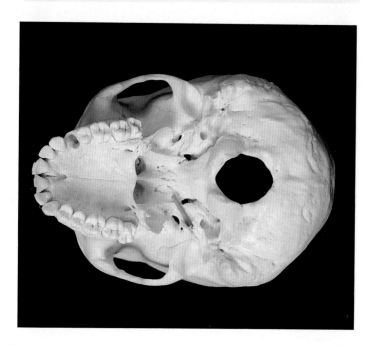

Some of the bones are more complex and will require special attention. The **temporal bone** has been the subject of a single dissection. It contributes to the middle and posterior cranial fossae, the lateral and inferior aspects of the skull. This bone can be studied in the intact skull with the aid of the atlas.

The sphenoid, ethmoid and palatine bones are more difficult to understand. The sphenoid and ethmoid bones contribute to the cranium and to parts of the facial skeleton. The palatine bone is located entirely within the facial skeleton where it contributes to several regions. Use the illustrations to help you locate these bones.

When all the apertures leading from the orbit to other regions have been studied, study the following:
○ the three cranial fossae,
○ the palate,
○ the infratemporal fossa,
○ the pterygopalatine fossa,
○ the base of the skull.

Locate the sphenoid bone (marked in red) in the anterior and middle cranial fossae, the orbit, the temporal fossa, the infratemporal fossa, the pharyngeal region and the pterygopalatine fossa (10).

Locate the ethmoid bone (marked in yellow) in the anterior cranial fossa, the nasal cavity where it contributes to the walls and septum, and in the orbits (11 & 12).

Locate the palatine bones (marked in blue) on each side of the palate, the nasal cavity, the pterygopalatine fossa, the orbit, and contributing to the pterygoid process of the sphenoid bone (13 & 14).

13

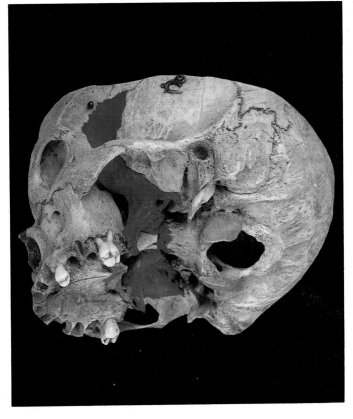

10

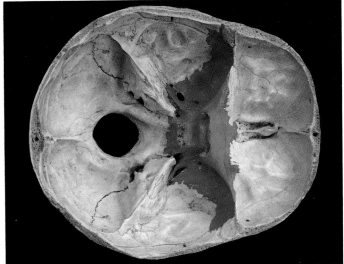
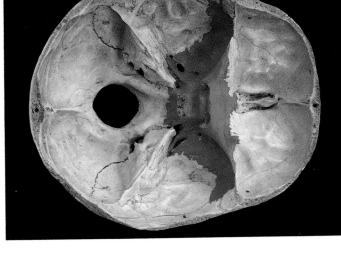

11

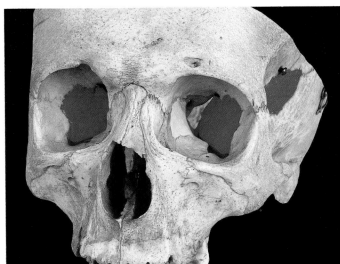
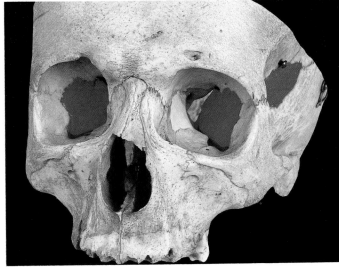

14

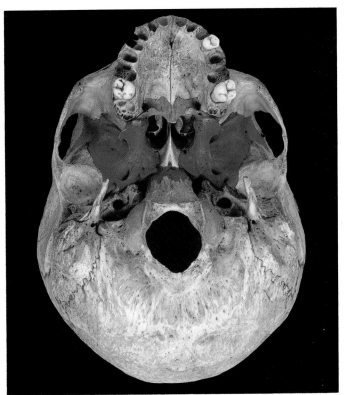

12

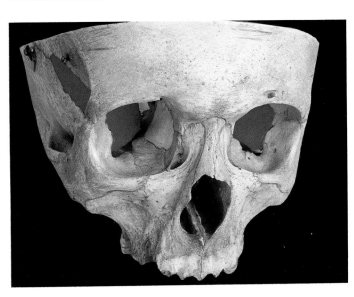

When this exercise is completed (repeat it several times), you can use the atlas to study the detailed anatomy of the separate bones. Any of the standard textbooks will help you to identify the structures that pass through the various foramina and fissures.

Surface Anatomy of the Head

Identify the visible structures on the head and face using the illustrations in the atlas (1):

1

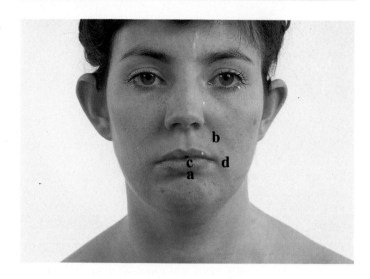

Eye

56 A

Sclera, iris, pupil, cornea, arcus senilis, eyelids, cilia.

Medial and lateral canthi, medial palpebral ligament palpebral fissure.

Lacus lacrimalis, lacrimal caruncle, plica semilunaris lacrimal papilla, lacrimal puncta.

Note the degree of protuberance of the eyeball.

Observe the rate of blinking. How long can you keep your eye open without blinking?

Ear

58 A

Auricle, external auditory meatus.
Observe the mobility of the outer part of the external auditory meatus.

Tragus, lobule (helix, scapha, tubercle, antihelix, antitragus, crus of helix, crus of the antihelix, intertragic notch, triangular fossa.

Nose

35 A

Root, bridge, apex, alae, nares.

Observe the mobile and fixed parts of the external nose, changes in dimensions of nares during inspiration and expiration.

Lips

35 A

*Philtrum, tubercle, labiomental groove (**a**), angle of the mouth, nasolabial sulcus (**b**), oral fissure (**c**).*

Observe the blanching of the lips in response to pressure.

Movements
Demonstrate protraction, retraction, elevation and depression of the mandible, and name the muscles producing these movements.

Demonstrate the movements of the eyes and name the extraocular muscles which produce these movements.

1. Demonstrate actions of the:
○ *orbicularis oris,*
○ *risorius,*
○ *zygomaticus major*
○ *mentalis,*
○ *occipitofrontalis,*
○ *palpebral and orbital parts of the orbicularis oculi.*

2. Demonstrate the following bony landmarks of the head: **9 A**
11 B
13 C
○ *nasion*
○ *inion (external occipital protuberance),*
○ *vertex,*
○ *supraciliary ridges, arches, glabella,*
○ *frontal tubers,*
○ *superior nuchal line,*
○ *temporal lines,*
○ *mastoid processes,*
○ *supramastoid crest,*
○ *parietal eminences,*
○ *zygomatic arch, tubercle of the zygomatic arch.*
○ *zygomatic bone,*
○ *suprameatal triangle,*
○ *suprameatal crest,*
○ *pre-auricular point,*
○ *condyle of the mandible with the mouth open and closed,*
○ *anterior border of the ramus of the mandible,*
○ *posterior border of the ramus of the mandible,*
○ *body of the mandible,*
○ *orbital margin,*
○ *supraorbital notch,*
○ *infraorbital foramen.*

3. Demonstrate the following palpable structures of the head and face:
○ *superficial temporal artery,*
○ *facial artery,*
○ *masseter muscle,*
○ *parotid duct,*
○ *modiolus (**d**),*
○ *mental foramen.*

4. Locate the following important structures with reference to visible and palpable landmarks:

Find the **pterion** on a dried skull and measure the distance from the frontal process of the zygomatic bone and the zygomatic arch using your fingers. You will probably find that the pterion is one thumb's breadth behind the frontal process and two fingers' breadth above the zygomatic arch.

Now palpate the bony landmarks in the living subject and use your fingers and thumb to find the approximate position of the pterion.

The **falx** and the **superior sagittal sinus** are below a line which connects the nasion with the inion across the vertex of the skull.

The **transverse sinus** lies deep to a line which would join the inion to the tragus of the auricle. When this line meets the back of the auricle, continue downwards where the auricle attaches to the head to the mastoid process to show the position of the sigmoid sinus.

The **anterior branch of the middle meningeal artery** inclines backwards from the pterion.

The **lower border of the cerebral hemisphere** can be represented on the side of the head by the line drawn for the transverse sinus above, which is continued along the zygomatic arch in front of the ear before turning up sharply to meet the pterion. The lower border continues along a line approximately 1 cm above the supraciliary ridges.

The main part of the **lateral fissure** runs posteriorly from the site of the pterion parallel to the zygomatic arch.

The **central sulcus** starts approximately 1 cm behind the mid-point of the line joining the nasion and inion. It runs inferoanteriorly towards the lateral fissure.

The **parotid gland** is located between the line drawn between the mandibular angle and mastoid process, and the zygomatic arch. It extends forwards for a variable distance over the masseter muscle.

The **parotid duct** can be rolled on the clenched masseter muscle just under 1 cm below the zygomatic arch.

To find the main branches of the **facial nerve**, the palm of the hand is placed over the position of the parotid gland with the fingers and thumbs spread out as far as possible and with the thumb pointing downwards:
○ the thumb now represents the cervical branch,
○ the index finger represents the marginal mandibular branch,
○ the middle finger represents the buccal branch,
○ the ring and little fingers represent the zygomatic branches.
Furthermore, the digits all point to the groups of muscles supplied by these branches.

5. **The apertures**
If ophthalmolscopes and other instruments are not available during the practical class, use the illustrations below. In any case, do not attempt to use instruments except under the supervision of a medically qualified member of staff.

Eye
Fornices of the conjunctiva, conjunctival sac, tarsal plates, tarsal glands (**2, 3, 4**).

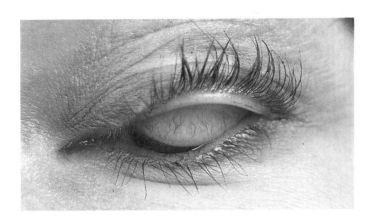

2

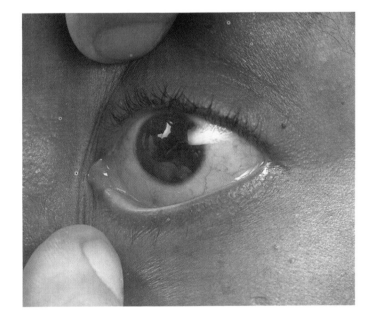

3

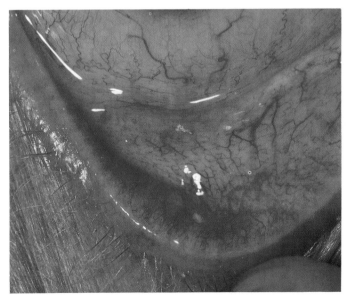

4

5

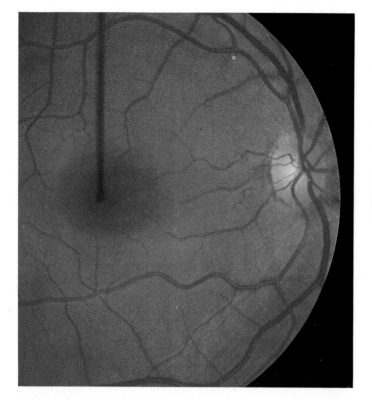

7a

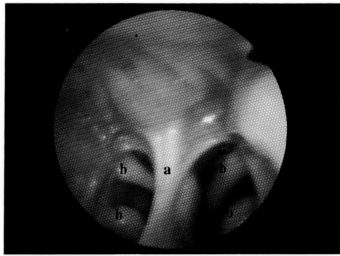

7b

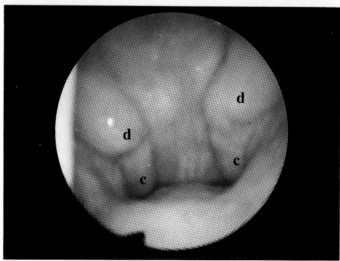

Ophthalmoscopy: macula and fovea centralis, optic disc, superior and inferior nasal, temporal and macular vessels (5).

Ear
External auditory meatus (6).
Tympanic membrane: flaccid part (a), anterosuperior and other quadrants.
Hande and lateral process of malleolus (b), long crus of incus, 'cone of light' (c).

Nose
Vestibule, septum.
Posterior rhinoscopy (7): septum (a), conchae (b), auditory tube (c) and tubal elevation (d).

Mouth
○ *Vestibule, orifice of parotid duct, frenula of lips, gums, teeth.*

○ *Tongue, frenulum, fimbriated fold, profunda linguae vein (8).*

6

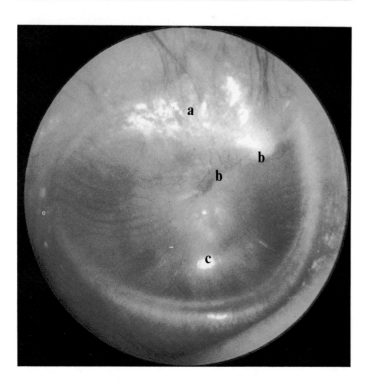

8

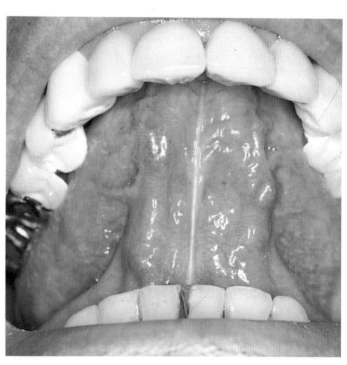

- ○ *Sulcus, sulcus terminalis, foramen caecum, vallate papillae, fungiform (red) papillae, filiform papillae, foliae linguae, sublingual gland, sublingual papilla (9).*

- ○ *Hard palate, transverse palatine folds, incisive papilla.*

- ○ *Soft palate, uvula, hamulus, soft palatoglossal and palatopharyngeal folds or arches, tonsils pterygomandibular raphe.*

- ○ *Palpate the lingual nerve as it passes close to the mandible near the third molar tooth, anterior margin of mandibular ramus.*

Larynx

- ○ *Laryngoscopy: median glossoepiglottic fold, epiglottic vallecula, epiglottis, epiglottic tubercle, aryepiglottic fold, vestibule, vestibular fold, vocal fold, vocal process (10).*

- ○ *Ventricle, cuneiform and corniculate tubercles, piriform recess.*

- ○ *Cuneiform and corniculate tubercles, piriform recess.*

9

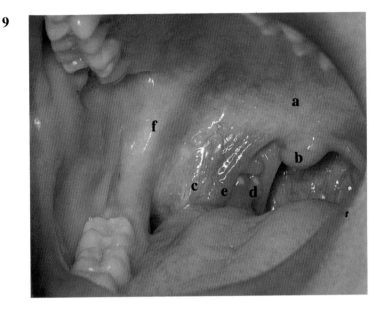

a. **Soft palate**
b. **Uvula**
c. **Soft palatoglossal fold or arch**
d. **Soft palatopharyngeal fold or arch**
e. **Tonsils**
f. **Pterygomandibular raphe**

10

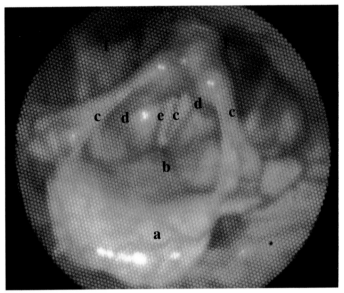

a. **Epiglottis**
b. **Epiglottic tubercle**
c. **Aryepiglottic fold**
d. **Vestibular fold**
e. **Vocal folds**
f. **Piriform recesses**

20

Review of the Head and Neck

The process of dissection has forced you to study the numerous separate structures in the head and neck, in artificial craniofacial regions. This review should help to draw together all these separate regions and their contents. You will need a dried skull as well as the specimens obtained from your dissections.

Look at the anterior aspect of the skull. Pass a bristle through the mandibular notch, the pterygomaxillary fissure and the sphenopalatine foramen. Look through the **piriform aperture** to see the tip of the bristle near the roof of the nasal cavity.

The bristle passes through the following regions:
○ the **parotid/masseteric region**,
○ the **infratemporal fossa**,
○ the **pterygopalatine fossa**,
○ the **nasal cavity**.

Now follow a real structure, the **ophthalmic division of the trigeminal nerve**, through several craniofacial regions. It starts in the **middle cranial fossa**, and passes forwards in the wall of the **cavernous sinus** to enter the **orbit** through the **superior orbital fissure**.

Before it enters the orbit, it divides into three branches: the **lacrimal branch** runs laterally to the lacrimal gland; the **frontal branch** runs close to the roof of the orbit, to the supratrochlear and supraorbital notches on the upper part of the face; the **nasociliary branch** runs close to the medial wall of the orbit.

A branch of the nasociliary nerve, the **anterior ethmoidal nerve**, runs through the anterior ethmoidal foramen to the **ethmoidal air cells**, back into the **cranial cavity**, and into the **nasal cavity** through the anterior part of the cribriform plate of the ethmoid bone. The position of this nerve is often marked by a groove on the deep surface of the nasal bone.

Follow the main trunk of the **mandibular division of the trigeminal nerve**, from the middle cranial fossa:
● in the wall of the **cavernous sinus**,
● through the **foramen rotundum**,
● across the top of the **pterygopalatine fossa**,
● through the **inferior orbital fissure**,
● in the **infraorbital groove and canal**,
● through the **infraorbital foramen** on the face.

45 C

Follow two of the main branches of the **mandibular division of the trigeminal nerve**. The mandibular nerve leaves the middle cranial fossa through the **foramen ovale**.

Trace the course of the **inferior alveolar nerve**, from the foramen ovale, across the **infratemporal fossa**, to the **mandibular foramen**.

Similarly, trace the **lingual nerve** from the foramen ovale to the medial surface of the body of the mandible, just inferior to the third molar tooth. (Both of these nerves branch off from the anterior division of the mandibular nerve, slightly below the foramen ovale, in the infratemporal fossa.)

The **facial nerve** has a complicated course through the temporal bone where several branches arise. Place a bristle in the **internal auditory meatus**; this is stopped where the facial nerve turns backwards at the **geniculate ganglion**.

Place another bristle in the **stylomastoid foramen** on the base of the skull. If this is pushed gently into the foramen, it may enter the **facial canal** inside the temporal bone; after passing the position of the geniculate ganglion, the bristle should enter the middle cranial fossa through the **hiatus for the greater petrosal nerve**. Follow the greater petrosal nerve from its entry into the middle cranial fossa to:
○ the **foramen lacerum** (it changes its name here),
○ through the **pterygoid canal**,
○ into the **pterygopalatine fossa**, where it ends in the **pterygopalatine ganglion**.

Observe the structures lying lateral to the pharynx(**1**). Note them in order, from the medial to the lateral position:
○ the **carotid sheath**,
○ the **styloid process** and its muscles,
○ the branches of the **external carotid artery**,
○ the **mandible** and **masseter muscle**.

50 A

43 B

Match the markings on the base of the skull to the diagram of a cross-section through the neck(**1**).

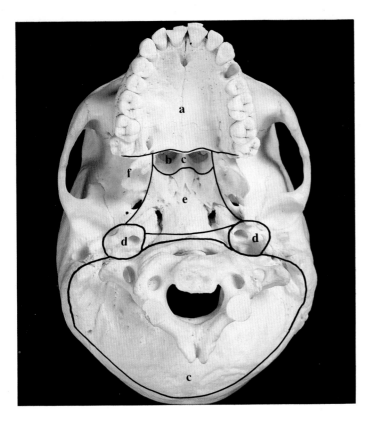

1a

1b

a. **Palate**
b. **Posterior apertures of the nose**
c. **Musculoskeletal compartment**
d. **Carotid sheath**
e. **Pharyngeal region**
f. **Infratemporal fossa**

2

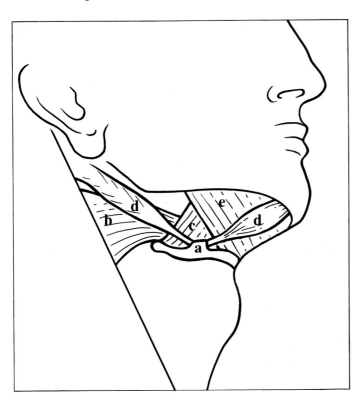

*Look in the region of the **hyoglossus muscle** (2) Note that the muscles are arranged in four layers:*
○ *the superficial, **digastric muscle**,*
○ *followed by the **mylohyoid muscle**,*
○ *followed by the **hyoglossus** itself,*
○ *and, deep to the hyoglossus muscle, the **middle constrictor of the pharynx**.*

Relate the following structures to the layers of the muscles:
○ *the **facial artery**,*
○ *the **submandibular salivary gland**,*
○ *the **hypoglossal nerve**,*
○ *the **lingual nerve**,*
○ *the **submandibular duct**,*
○ *the **stylohyoid ligament**,*
○ *the **lingual artery**.*

43 B

44 A

a. **Hyoid bone**
b. **Middle constrictor**
c. **Hyoglossus**
d. **Digastric**
e. **Mylohyoid**

Upper Limb

21

The Scapular Region

The purpose of this dissection is to demonstrate the muscles of the back which are attached to the scapula, and the nerves which supply these muscles.

The muscles of the scapular region may be considered in two groups: (a) those moving the scapula on the thoracic wall (trapezius, levator scapulae, the rhomboids); (b) those acting at the glenohumeral joint (latissimus dorsi, teres major).

Place the cadaver in a prone position.

Before starting the dissection, identify the following bony landmarks on the cadaver:
the occipital bone, the mastoid process, the spines of the seventh cervical and first thoracic vertebrae, the spine and acromion of the scapula and iliac crest.

Make the following incisions in the skin (1):
- in the midline, from the occipital bone to the tip of the coccyx,
- along the iliac crest,
- from the mastoid process laterally to a point halfway down the lateral surface of the upper arm;
- make two other incisions from the midline extending to the mid-axillary line.

Reflect the skin carefully to form three skin flaps. The connective tissue of the back is quite tough in places, tougher than the muscles which will be damaged if you try to remove the skin and connective tissue together (2).

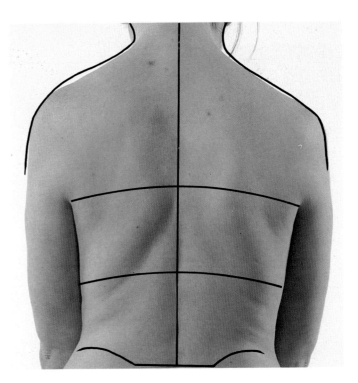

1

2

3

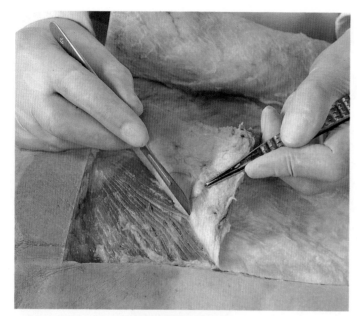

Remove the superficial fascia and the deep fascia of the back and the posterior surface of the arm piecemeal. Identify any nerves that you find, but do not spend time searching for them (**3**).

Clean the edges of the *trapezius* and *latissimus dorsi muscles*. Determine the directions taken by the fibres of these muscles, and develop the fascial plane beneath them using your fingers. Define the origins and insertions of these muscles (**4 & 5**).

87

> *Study the dissection at this point. Observe the directions taken by the fibres in the various parts of the trapezius muscle, check the origin and insertions of this muscle and map out the muscle on an articulated skeleton.*
>
> *Confirm that the latissimus dorsi muscle arises from the thoracolumbar fascia and iliac crest, and that its tendon forms part of the posterior axillary fold as it approaches its insertion on the humerus.*
>
> *Find the* **lumbar triangle**, *between the free edges of the latissimus dorsi and external oblique muscles and the iliac crest, and the* **triangle of auscultation** *between the upper free edge of latissimus dorsi, trapezius and the medial border of the scapula.*

111 B

4

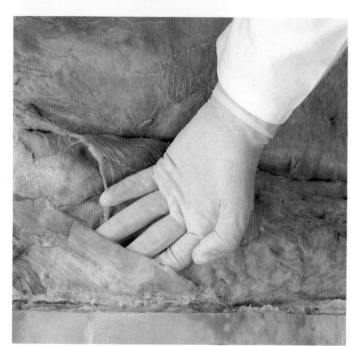

Divide the trapezius muscle as close as possible to its origins on the spines of the cervical and thoracic vertebrae (**6**). Develop the planes between the muscles attached to the medial border of the scapula, (*levator scapulae, rhomboideus minor* and *rhomboideus major*) and those attached to the lateral border (*teres major* and *teres minor* (**7**)).

5

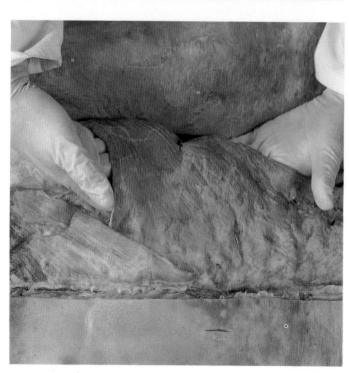

Study the dissection again. Note the directions taken by the fibres of the muscles you have displayed, and their origins and insertions.

112 A

111 B

Note that the posterior axillary fold contains the tendon of latissimus dorsi winding around the bulk of the teres major muscle (7).

Find the accessory nerve on the deep surface of the trapezius muscle (8).

Try to find the dorsal scapular nerve and artery lying deep to the rhomboid muscles, close to the medial border of the scapula.

If the head and neck have been dissected, trace the levator scapulae muscle and the accessory nerve into the neck.

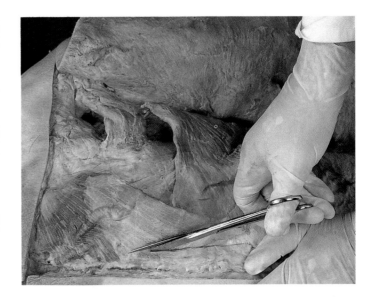

6

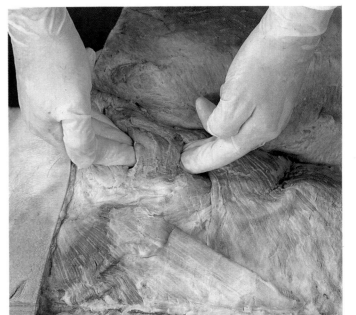

7

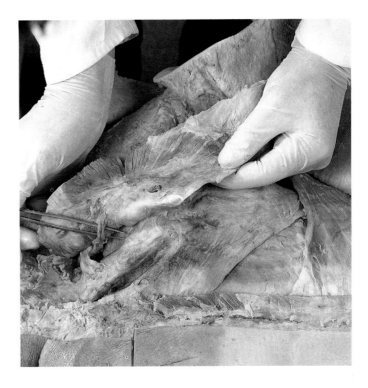

8

The Deltoid Region

The purpose of this dissection is to study the deltoid muscle and then to reflect it to see the muscles which lie deep to it and to find the axillary nerve.

1

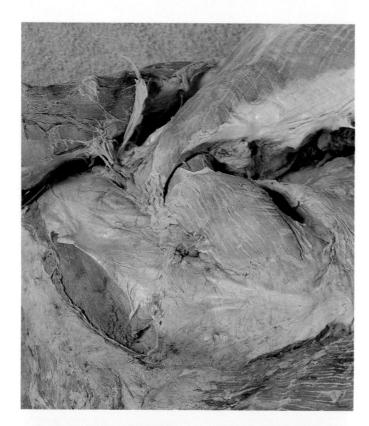

Keep the cadaver in the prone position and abduct the arm slightly.

Clean the *deltoid muscle* thoroughly.　　**113 B**

> *Note the origins and insertion of the deltoid muscle and the directions taken by its fibres.*

Detach the deltoid muscle from its scapular origin, but leave it attached to the clavicle.

> *Note the **axillary nerve** and the **circumflex humeral vessels** entering the deep surface of the muscle (**1**).*

Remove the deep fascia and clean the *supraspinatus* and *infraspinatus muscles* on the dorsal surface of the scapula (**2**).

> *Study the origins and insertions of the **supraspinatus**,*　　**113 C**
> ***infraspinatus** and **teres minor** muscles.*
>
> *Note that the **acromion** forms an arch over the tendon of supraspinatus and that the tendons of the three muscles blend with the **capsule of the shoulder joint**.*
>
> *Confirm that the axillary nerve supplies both the deltoid and teres minor muscles.*

2

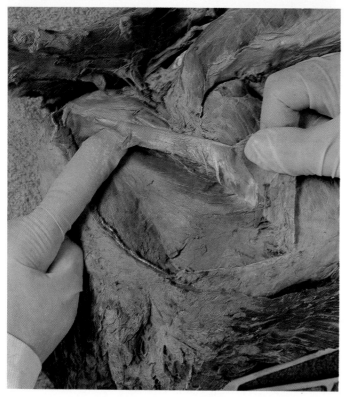

Divide the supraspinatus muscle as far medially as possible and reflect it laterally.

> *Find the **suprascapular nerve** and trace it into the*　　**113 C**
> *supraspinatus, infraspinatus and teres major.*
>
> *Find as many as possible of the arteries which contribute to the collateral circulation around the scapula.*

23

The Breast

The aim of this dissection is to display the relations of the female breast. If you have a male cadaver, follow the instructions as a preliminary to dissecting the pectoral region.

Place the cadaver in the supine position.

Make the incisions shown in the illustration, carefully encircling the nipple (**1**).

Reflect the skin. Look for the platysma and the supraclavicular nerves (**2**).

> *Note the extent of the breast, that is, which ribs and intercostal space does it cover? Abduct the arm to 90° and find the axillary tail.*

Lift the breast and superficial fascia away from the deep fascia by developing the plane between them with your fingers (**3**)

Further dissection of the breast is unlikely to reveal anything of its structure or blood supply. Students are advised to read an account of the anatomy of the breast in a standard textbook.

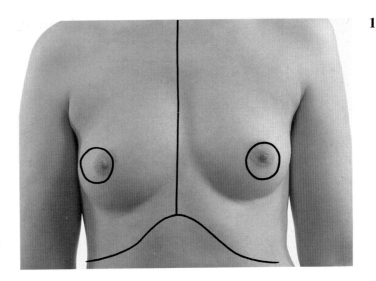

1

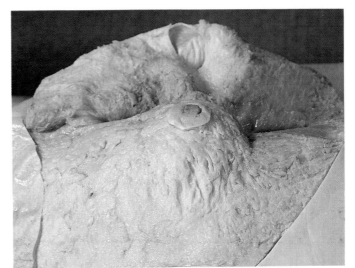

2

3

The Pectoral Region

*The purpose of this dissection is to display the muscles of the anterior
axillary fold and their nerve supply.*

1

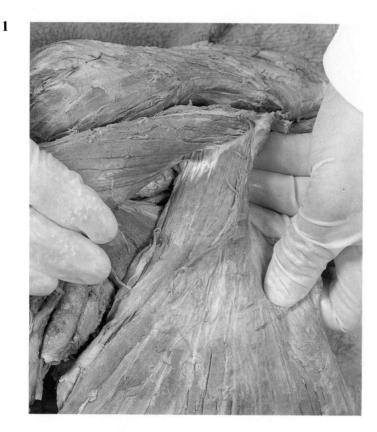

2

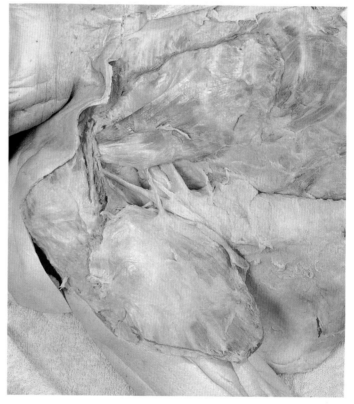

Clean the *pectoralis major muscle* as far laterally as its
insertion into the humerus.

Divide the pectoralis major as close as possible to its
origins and reflect it laterally, taking care to preserve
the vessels and nerves entering its deep surface (**2**).

108 A   *Observe the origins of the **pectoralis major muscle**,
and the way in which its sternocostal and clavicular
parts insert into the humerus (**1**).*

3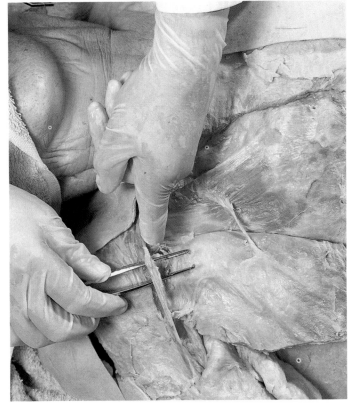

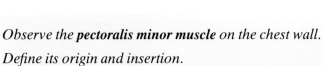

4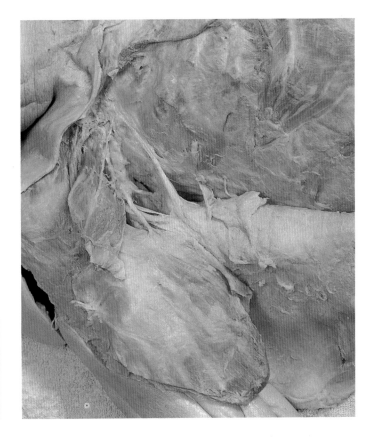

109 B *Observe the **pectoralis minor muscle** on the chest wall.*

Define its origin and insertion.

*Trace the **cephalic vein** between the deltoid and pectoralis major to the point where it pierces the **clavipectoral fascia** (between the clavicle and pectoralis minor) (**3**).*

Detach the pectoralis minor from the thoracic wall, close to its origin, and reflect it upwards, taking care to preserve the nerve and vessels entering its deep surface (**4**).

*Identify the **medial and lateral pectoral nerves** and the arteries which pierce the clavipectoral fascia.* 109 B

25

The Axilla

The walls of the axilla have been seen in earlier dissections. These will be studied again because it is necessary to understand the shape of the axilla in order to appreciate the positions of important groups of lymph nodes. The lymph nodes are not easy to find by dissection.

The fascial planes will then be developed to demonstrate the axillary sheath. The axillary sheath will be opened to show the cords of the brachial plexus surrounding the axillary artery.

Next, the five main branches of the brachial plexus will be demonstrated, as well as the branches of the axillary artery.

Finally, if the neck has been dissected, the brachial plexus will be traced upwards to its trunks and roots in the neck.

1

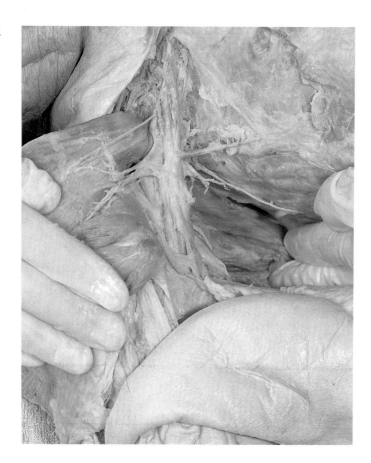

2

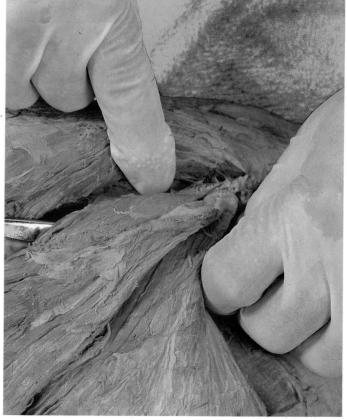

116 A,B

Review the structure of the posterior axillary fold, eres teres major and latissimus dorsi (1).

Review the structure of the anterior axillary fold, mainly pectoralis major (2).

Now clean the medial wall of the axilla to find the **serratus anterior muscle**.

Push your fingers between the serratus anterior and scapula, and find the **long thoracic nerve** *on the serratus anterior in the mid-axillary line.*

Verify that the axilla presents as a pyramidal space between the anterior and posterior axillary folds and the upper part of the thoracic wall.

Abduct the arm to the horizontal position. Use your fingers to develop the fascial planes around the *axillary sheath* (3).

Use scissors to open the axillary sheath. Separate the structures in the sheath by blunt dissection, using blunt forceps.

Cut away all or most of the veins in the axillary sheath (**4**).

Study the three cords of the brachial plexus related to the axillary artery. **117 C**
118 A

Find the five major branches of the plexus, and note their origins from the cords (**5**).

Trace the medial and lateral pectoral nerves from the plexus to the pectoral muscles.

If the neck has been dissected, trace the whole brachial plexus from its roots to its terminal branches in the axilla (**6**).

Return pectoralis minor to its usual position to identify the three parts of the axillary artery. Find as many branches of the artery as you can.

3

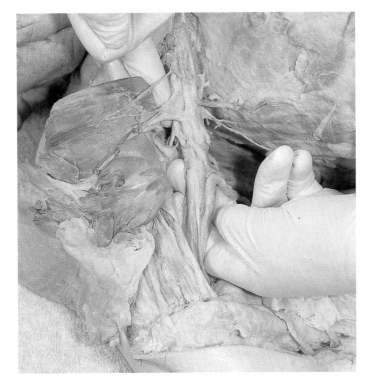

4

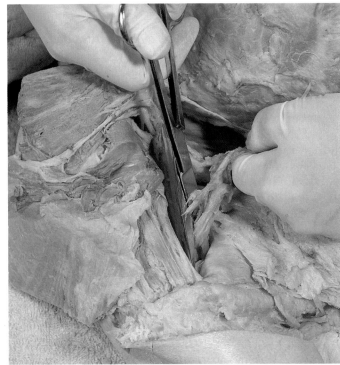

5

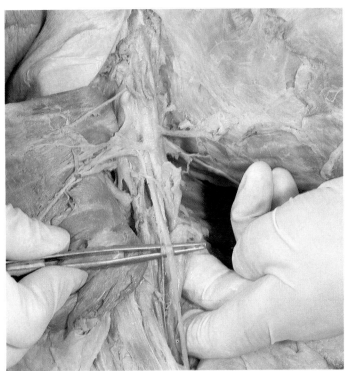

6

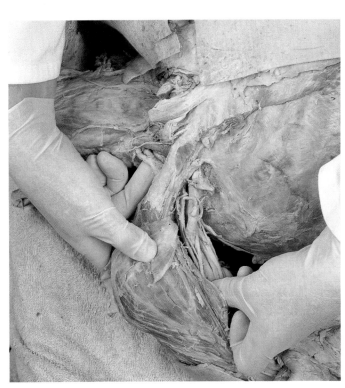

26

The Back

This is a difficult and time-consuming dissection. It is described in three parts: firstly, the upper limbs are separated from the trunk; secondly, the lower part of the back is dissected; thirdly, the cervical part of the vertebral column is dissected.

Divide the *levator scapulae* and *rhomboid muscles* as close as possible to the origins on the vertebral column (**1**).

Detach the slips of the *serratus anterior* from the ribs.

Divide the *brachial plexus* and *axillary vessels* as high up as possible in the axilla.

Pass your hand between the serratus anterior and the wall of the thorax, to separate the upper limb from the thorax. Put the isolated upper limb on one side (**2**).

87 Cut through the *thoracolumbar fascia* close to the spines of the lower vertebrae, and reflect this and the *latissimus dorsi muscle* laterally. Do this on both sides.

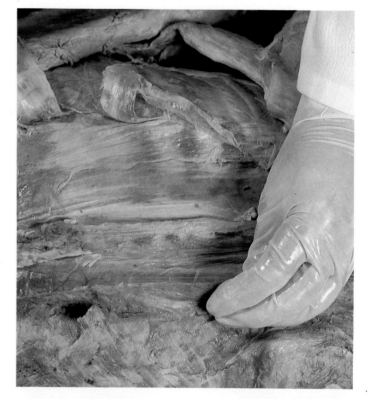

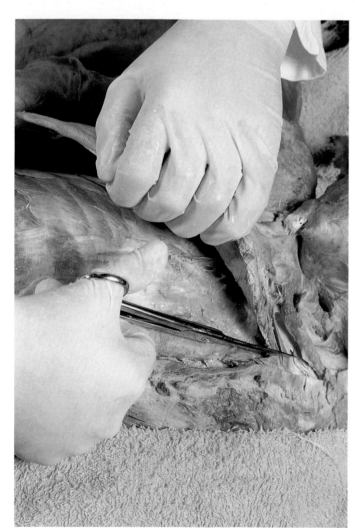

88 A

Observe the *erector spinae muscles*. Do not try to separate them into their component parts. Remove them by cutting through their attachments with scissors (**3 & 4**).

Observe the transverse processes, laminae and spines of the lower vertebrae. Saw through the laminae of the vertebrae in the lumbar and lower thoracic regions, and remove the section (**5**).

Identify and remove the *venous plexus* surrounding the dural sheath of the spinal cord. Using fine scissors, make a longitudinal incision in the *dura* and the underlying *arachnoid* (**6**).

84 A,B

86 B,C

*Open the dura and observe the **spinal cord** and **nerve roots** within the dural sheath.*
*Find the **filum terminale** and **conus medullaris**. Note the vertebral level of the tip of the conus medullaris (**7**). Identify the dorsal and ventral nerve roots, and note the obliquity of their passage through the dural sheath to the appropriate intervertebral foramen (**8**).*

4

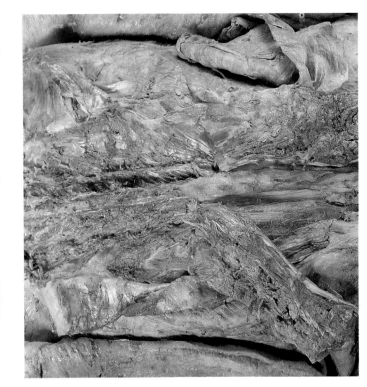

5

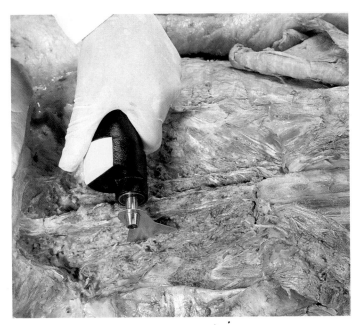

6

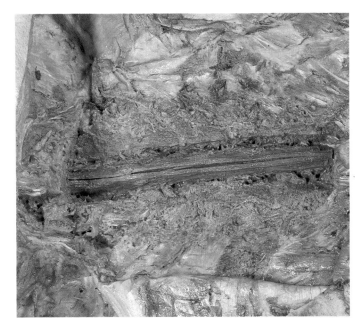

7

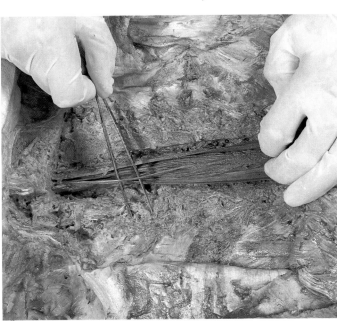

8

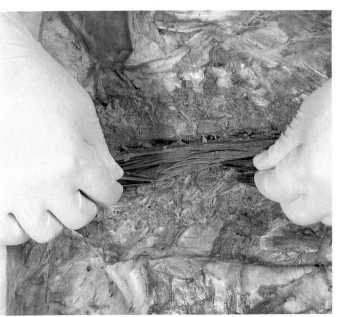

Use ronguers to nibble away the bone in the region of one or two intervertebral foramina.

*Observe a **dorsal root ganglion** as well as the way in which a dorsal and ventral root join together to form a **mixed spinal nerve**.*

Dissection of the cervical region of the vertebral column
This belongs with the dissection of the head and neck, but it is more convenient to include it with the dissection of the back.

Before starting to dissect, look at the upper thoracic and cervical vertebral column, and the base of the skull, on an articulated skeleton.

Reflect both halves of the trapezius muscle laterally, if this has not yet been done.

*Observe the **splenius capitis** and **semispinalis muscles** (9).*

Remove the splenius and semispinalis muscles, by detaching them from their insertions on the base of the skull and reflecting them downwards.

Pick out the muscles and connective tissue in the suboccipital region. Try to locate the *occipital artery* and *great occipital nerve*; do not attempt to identify the small muscles, some of which were damaged during the exposure of the back of the pharynx

Observe the laminae and spines of the cervical vertebral column (10). Divide the laminae of the *atlas*, *axis*, and one or two typical cervical vertebrae (11). Open the dura, just as you did in the lumbar region.

83 D,E,F   *Observe the cervical part of the spinal cord and **dorsal roots of the cervical nerves** (12).*

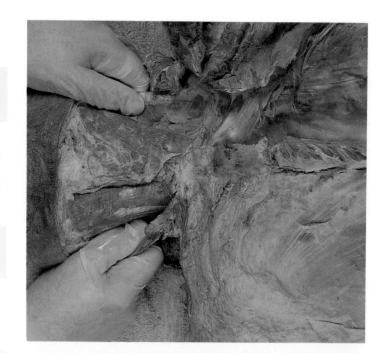

9

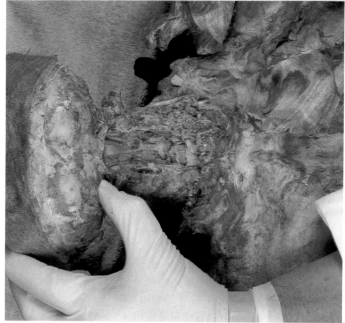

10

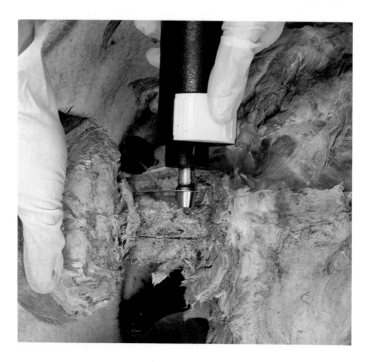

11

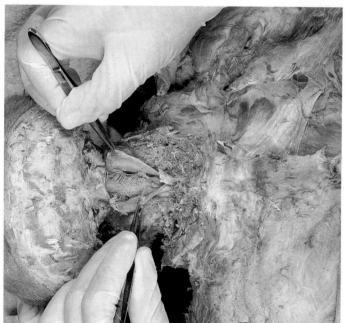

12

Remove the cervical part of the dura and spinal cord, and observe the *posterior longitudinal ligament*.

Remove the posterior longitudinal ligament to observe the ligaments around the atlanto-occipital and atlantoaxial joints (**13**).

13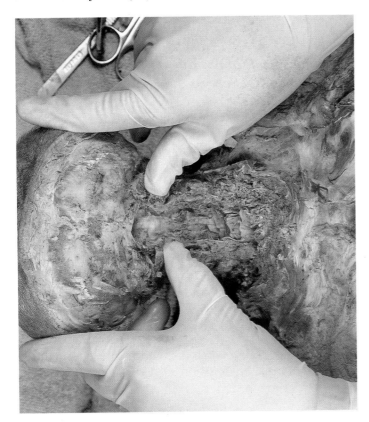

The Arm

In this dissection the skin and deep fascia are removed to display the muscles of the anterior and posterior compartments of the arm. The musculocutaneous, median, ulnar, axillary and radial nerves, and the brachial artery, will then be traced through the arm.

1

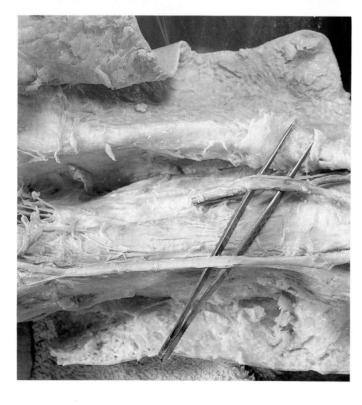

Make the incisions as shown in the illustration. Take care to avoid damage to the structures in the *cubital fossa* (**1**).

Note the arrangement of veins in the arm and cubital fossa, and compare your specimen with the illustrations in the standard textbooks of anatomy.

124 A, B

Remove the superficial and deep fascia from the arm, but preserve the *cephalic*, *basilic* and *median cubital veins* (**2**).

Develop the fascial planes around the muscles in the anterior compartment of the arm (**3**).

*Observe the **biceps** and **coracobrachialis muscles**. Follow the median nerve as far as the cubital fossa, and note its relation to the brachial artery throughout the arm.*

119 B 124 B, C

Follow the musculocutaneous nerve into the muscles of the front of the arm and identify its cutaneous branch.

2

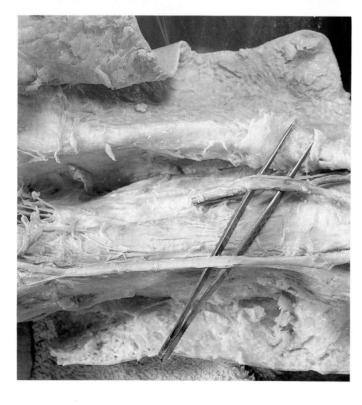

3

Develop the fascial planes around the three parts of the triceps muscle in the posterior compartment of the arm. Develop the quadrangular and triangular spaces of the arm.

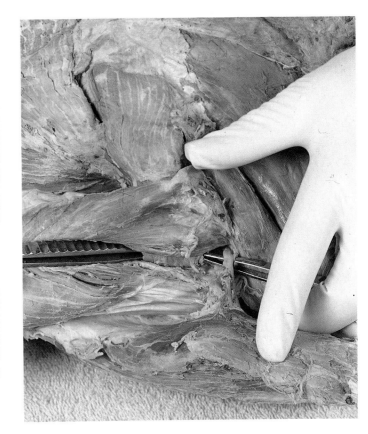

*Observe the **axillary nerve** passing through the quadrangular space. Trace it back to the posterior cord of the brachial plexus (**4**).*

*Look through the triangular space to see the **radial nerve** running in the **spiral groove** of the humerus (**5**).*

Divide the lateral head of the triceps in order to see the radial nerve more clearly (**6**).

Observe the radial nerve in the spiral groove.

*Follow the **ulnar nerve** from the brachial plexus through the medial intermuscular septum, into the posterior compartment of the arm.*

The flexor compartment of the forearm will be dissected before the full dissection of the cubital fossa.

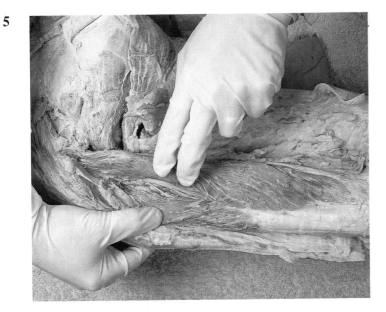

The Flexor Compartment of the Forearm

When the skin and fascia have been removed, the muscles of the flexor compartment of the forearm will be studied in three layers: superficial, intermediate and deep. Then the radial, median and ulnar nerves and the arteries of the forearm will be found, and their relations to the three layers of muscles will be noted.

1

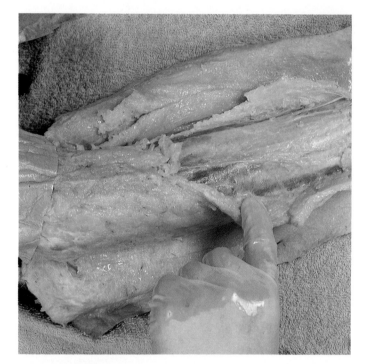

Note that the illustrations in standard textbooks show the structures of the forearm in a fully supinated anatomical position. The forearm in your specimen is likely to be in the mid-prone position, so the location of the anterior compartment should be carefully noted before you start to dissect.

Make an incision in the skin overlying the anterior compartment from the cubital fossa to the wrist. Take care to avoid damage to the underlying structures, especially at the wrist where the skin is very thin.

Observe the veins in the superficial fascia.

Remove the superficial fascia and most of the veins (**1**). Leave the veins in the region of the cubital fossa.

Open the anterior compartment of the arm with a longitudinal incision. Use scissors to avoid damage to the muscles in the compartment.

2

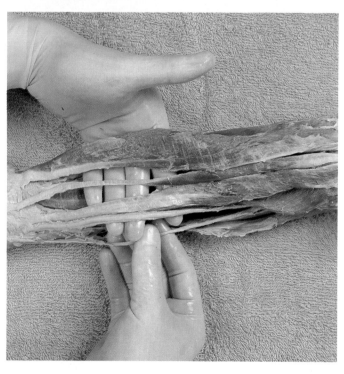

3

Turn back flaps of deep fascia to make the contents more visible.

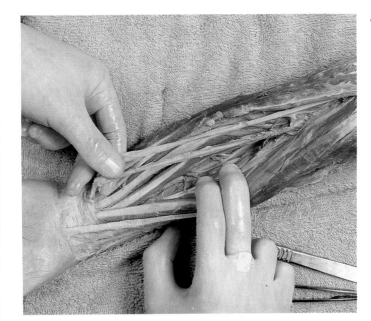

4

*Observe four muscles in the superficial layer: **pronator teres, flexor carpi radialis, palmaris longus** (this one may be absent) and **flexor carpi ulnaris**(2).*

*Observe the muscle of the intermediate layer **flexor digitorum superficialis**(3)*

126 A

*Push the muscles of the superficial and intermediate layers aside, so that the deepest muscles become visible: the **flexor pollicis longus, flexor digitorum profundus** and **pronator quadratus**(4).*

Identify the median nerve, and the radial and ulnar nerves and arteries close to the wrist. Trace them upwards towards the cubital fossa, noting their relations in the forearm.

29

The Cubital Fossa and some relations of the Elbow Joint

The arm and forearm have been dissected, and the positions of nerves and arteries in those regions have been noted in the last two sections. Here, the boundaries of the cubital fossa are demonstrated, and the median, radial and ulnar nerves are traced from the arm into the forearm.

1

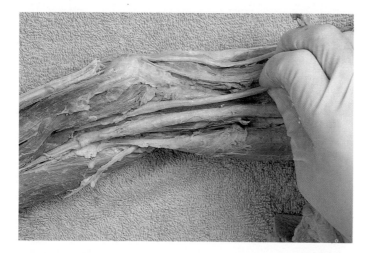

2

3

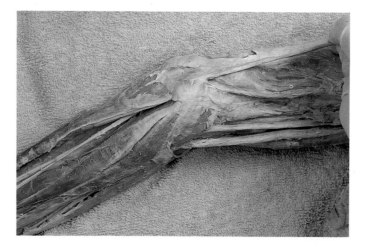

If necessary, clean the area around the cubital fossa by blunt dissection, using your fingers or closed blunt forceps.

Use the illustrations in the atlas to recognize the borders and contents of the cubital fossa. Pay particular attention to the structures lying beneath the **median cubital vein** *and those beneath the* **bicipital aponeurosis***. Find the insertion of the tendon of the biceps muscle.*

124 C

125 A, B

126 B

Trace the **median (1)**, **radial (2)** *and* **ulnar nerves (3)***, from the arm into the forearm.*

129 D

Pay particular attention to the relations of each nerve in the lower part of the arm, the upper part of the forearm, and around the elbow joint.

30

The Palm of the Hand and Carpal Tunnel

It is important to be able to recognize and name the nerves, vessels and tendons as they pass from the forearm into the hand. The palm of the hand is usually dissected in several layers, which is time consuming and not especially rewarding; therefore a simplified dissection is presented here in which the deeper structures are identified as the superficial ones are moved aside.

Make the incisions as shown in the illustration (**1**). Reflect the skin carefully and remove the superficial fascia.

Clean the muscles of the *thenar* and *hypothenar eminences*. Separate the muscles by blunt dissection, using the tips of closed forceps.

131 D
132 A
*Identify the muscles of the thenar and hypothenar eminences. Note the distal attachments of the **palmar aponeurosis**.*

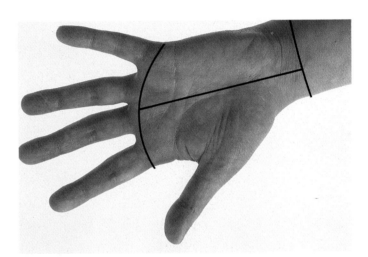

*Identify the bony points to which the **flexor retinaculum** is attached by palpation:*
- *the **hook of the hamate bone**,*
- *the **pisiform bone**,*
- *the **ridge of the trapezium**,*
- *the **tubercle of the scaphoid**.*

*Identify the branches of the **median** and **ulnar nerves** and the **ulnar artery** as they pass superficially to the flexor retinaculum.*

*Find the **radial artery** as it passes into the thenar eminence.*

Dissect off the palmar aponeurosis, but leave the tendon of ***palmaris longus*** (**2**).

Separate the structures in the hand by blunt dissection, using closed forceps (**3**).

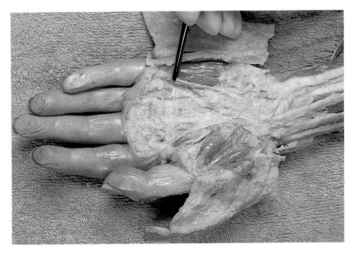

Find:
133 A
- *the **flexor tendons**,*
- *the branches of the **median** and **ulnar nerves**,*
- *the **superficial palmar arch**.*

132 A
*Push these structures aside to see the **lumbrical** and **palmar interosseous muscles**.*

Define the flexor retinaculum again. Pass a blunt
133 A
*probe deep to it, to demonstrate the **carpal tunnel**.*

Divide the flexor retinaculum as shown in the illustration to open the carpal tunnel (**4**).

Identify the median nerve and the long tendons as they pass through the carpal tunnel, and note their relations to each other.

The structures in the palm of the hand may now be moved aside easily, therefore examine the lumbrical and palmar interosseous muscles again.

126 A

*Move the tendons in the distal part of the forearm, to see the **pronator quadratus muscle**.*

31

The Extensor Region of the Forearm and Dorsum of the Hand

After removal of the skin and superficial fascia, the muscles of the extensor compartment will be considered in three groups: lateral, superficial and deep. The examination of the supinator muscle will be delayed until the joints are dissected.

Remove the remains of the skin, deep fascia and superficial fascia. Use your fingers to develop the muscles of the lateral and superficial groups.

*Grasp the three muscles of the lateral compartment, **brachioradialis**, **extensor carpi radialis longus** and **extensor carpi radialis brevis** between your thumb and forefinger. Note their approximate positions of their origins and insertions (**1**).*

127 C
*In a similar way separate the muscles of the superficial group and identify **extensor digitorum**, **extensor digiti minimi** and **extensor carpi ulnaris**. Define their origins and the insertion of extensor carpi radialis (**2**).*

127 D
*Move the superficial muscles to one side and identify the muscles of the deep group, especially those to the thumb (**3**)*
○ *the **abductor pollicis longus**,*
○ *the **extensor pollicis brevis**,*
○ *the **extensor pollicis longus**.*
*Try to find the **extensor indicis**.*

Clean the tendons of the deep group muscles to the thumb and trace them to their insertions.

139 C
132 B
143 B
Note the muscle insertions and boundaries of the 'anatomical snuff box'.

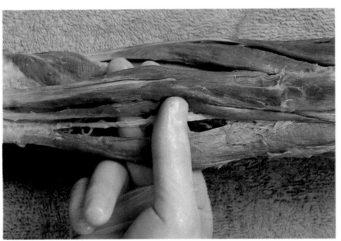

32

Dissection of the Finger

As injuries to the fingers are quite common, it is a good idea to study the anatomy of the finger in some detail.

On the palmar surface of the finger, reflect the skin and open the fibrous tendon sheath (**1**).

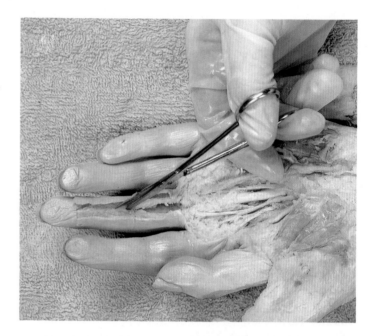

134 A
131 D
133 A,B

Follow the nerves, arteries and tendons into the finger, and observe the tendon insertions.

Similarly, remove the skin from the dorsum of the hand (**2**).

127 C

Observe the extensor expansion.

Look again at the lumbrical and interosseous muscles in your dissection.

137 D

Use the illustrations in the atlas to demonstrate the arrangements of the tendons and intrinsic muscles of the hand.

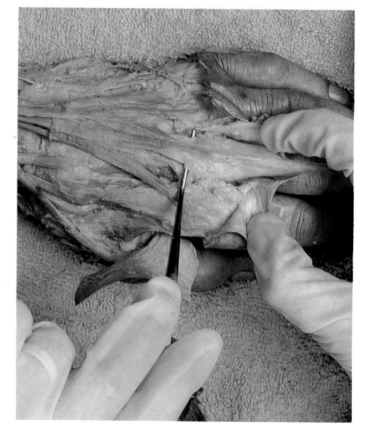

33

The Shoulder Joint

Use this dissection to revise the muscles of the 'rotator cuff', and then open the capsule of the joint to inspect the interior.

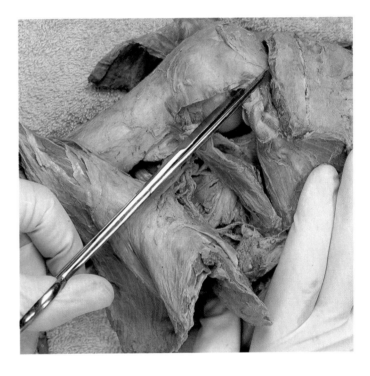

Identify the muscles of the rotator cuff around the shoulder joint:

112 A
○ *the* **supraspinatus,**
○ *the* **infraspinatus,**

113 B
○ *the* **teres minor,**
○ *the* **subscapularis,**
○ *the* **long head of the biceps** *muscle.*

Use scissors to cut through the capsule of the joint where it blends with the tendons of the rotator cuff muscles. Divide the long head of the biceps and open the joint (**1**).

114 D
Observe the articular surfaces and other areas, using the illustrations in the atlas.

Note especially where the capsule of the shoulder joint is loose and thin, and where it is reinforced by tendons as they go on to insert in the humerus.

34

The Elbow and Wrist Joints

*This is a destructive dissection, therefore choose the least well-dissected
of the two upper limbs. The muscles will be stripped off systematically
and the joint cavities will be opened.*

Expose the capsule of the elbow joint by cutting
through all the muscles which surround it, as close as
possible to their origins or insertions (**1**).

1

129 D,E
> *Use this opportunity to study the origins and
> insertions of these muscles, and to study the supinator
> muscle.*

**122 A,B,
C,D,E,F**
Clean the capsule of the joint to display the ligaments
(**2**).

Open the joint through the anterior part of the capsule
(**3**).

123 B
> *Use the illustration in the atlas to identify the various
> structures within the joint capsule.*

Cut through the flexor and extensor tendons of the
forearm just distal to the wrist joint.

2

**127 C,D
131 D
132 A**
> *Use this opportunity to revise the relations of the wrist
> joint. Identify the ligaments.*

Cut through the capsule of the joint, leaving only the
medial ligament to hold the two parts together (**4**).

**145 D,E
144 A,B,C**
> *Use the atlas to help you to identify the intra-articular
> structures.*

3

4

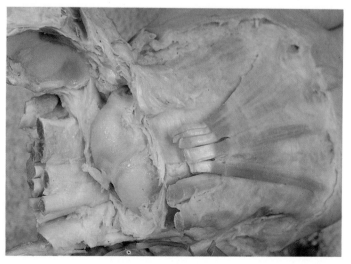

35

Osteology of the Upper Limb

You will need a set of bones of the upper limb, a fully articulated skeleton, a colleague to assist in demonstrating the bony landmarks in the living subject, and as many upper limb radiographs as possible.

92-102 *Use the illustrations in the atlas to identify the parts and markings on each bone.*

Learn all aspects of the following bones:
- *scapula,*
- *clavicle,*
- *humerus,*
- *radius,*
- *ulna.*

You should be able to define their anterior, posterior, medial and lateral surfaces or borders, and to distinguish the superior end of each bone from its inferior end. You will then be able to determine the side of the body to which each of these bones belongs, holding it in its correct anatomical position. You may check your assessment against the articulated skeleton.

Review the attachments of muscles. Those will have already been observed in the respective dissections and examinations of the joints. If necessary, use the illustrations in the atlas to refresh your memory. Do not plot out the origins and insertions of the muscles on isolated bones; instead, use the articulated skeleton to understand and remember the actions of each muscle.

With respect to each bone, note the following:

Scapula

Which vessel or nerve passes through the suprascapular notch?

The scapula is usually wired on to an articulated skeleton, you will therefore need to study its movements on a living subject. Define by palpation the medial border and spine, as well as the inferior angle of the scapula.

Note the levels of the angles with respect to the ribs and vertebral column in the articulated skeleton. Note the change in position of these landmarks during:
- *protraction of the scapula,*
- *retraction of the scapula,*
- *shrugging of the shoulders,*
- *full abduction of the arm.*

Clavicle

Plot out the attachments of the ligaments to the clavicle and adjacent structures:
- *costoclavicular ligament,*
- *coracoclavicular ligament (trapezoid and conoid parts).*

Observe the sternoclavicular joint in the articulated skeleton and living subject.

Humerus
Note the relations of:
- *the **posterior circumflex humeral artery** and **axillary nerve** to the **surgical neck of the humerus** (a),*
- *the **profunda brachii artery** and **radial nerve** to the **spiral groove** on the posterior aspect of the humerus (b),*
- *the **ulnar nerve** to the **medial epicondyle** of the humerus (c).*

*Assemble the radius, ulna and humerus, as shown in the atlas. Note the **carrying angle**, most obvious in the female (2).* **101 C,D**

Note the arrangement of the two epicondyles of the humerus and olecranon in the living subject: **121 D**
- *with the forearm extended at the elbow joint,* **122 C**
- *with the elbow flexed.*

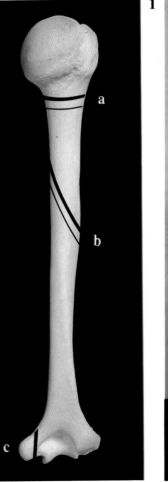

1

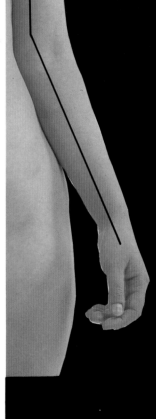

2

*Find the **head of the radius** on the posterior surface of the forearm, just distal to lateral epicondyle of the humerus. Feel the head of the humerus rotating as the subject pronates and supinates his or her forearm.*

Observe the movement of the radius around the ulna in pronation both in the articulated skeleton and in the living subject.

Find the following structures of the wrist in the articulated skeleton and in the living subject:
- *the styloid process of the ulna (3 & 4),*
- *the styloid process of the radius (3 & 4),*
- *the dorsal tubercle of the radius (Lister's tubercle) (3 & 4),*
- *the pisiform bone* } *at the distal skin*
- *the tubercle of the scaphoid* } *crease (5 & 6),*
- *the hook of the hamate* } *in the hypothenar and*
- *the tubercle of the* } *thenar eminences*
- *trapezium* } *respectively (5 & 6).*

Use the articulated skeleton to determine the way the proximal row of carpal bones articulates with the radius and ulna. Use radiographs to show how these bones move with respect to each other during radial and ulnar deviations of the wrist.

146 B
145 F

*Study the **carpal bones** and learn to identify them in radiographs.*

146 A

With the aid of the atlas, study ossification of the bones of the upper limb; use radiographs to study ossification of the carpals of the hand in different age groups.

106

3

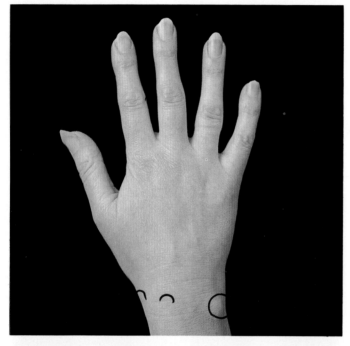

4

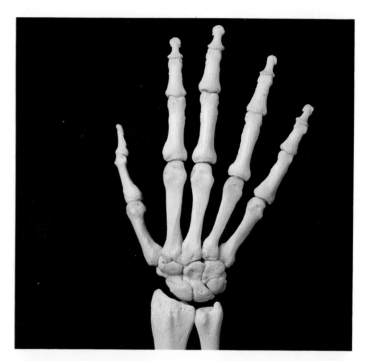

5

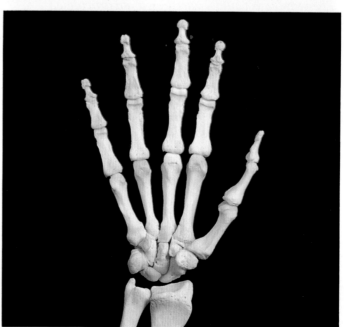

6

36

Surface Anatomy of the Upper Limb

It is important that you learn to identify landmarks, feel pulses and observe actions of muscles on a living subject other than yourself.

To demonstrate movement at a joint, observe the subject while he or she performs that movement starting from the anatomical position. Note the range of each movement.

To identify the muscle or muscles involved in a particular movement, ask the subject to perform it against resistance (that is, try to block the movement), so that you can see or feel the contraction of the muscles.

107

110 A

The Shoulder
Palpate the following bony landmarks:
○ the **coracoid process** of the scapula (deep in the **deltopectoral triangle**),
○ the other landmarks listed in the previous section,
○ the entire clavicle.

Observe the following contours:
○ the **deltoid muscle**, overlying the greater humeral tuberosity,
○ the **deltopectoral triangle**,
○ the **anterior** and **posterior axillary folds**,
○ the **serratus anterior** muscle.

Find out which muscles are active in the following movements of the scapula on the thoracic wall:
○ elevation,
○ protraction,
○ retraction,
○ rotation.

Demonstrate the following movements at the glenohumeral joint:
○ flexion,
○ extension,
○ abduction (observe the movement of the scapula during full abduction),
○ adduction,
○ medial and lateral rotation (with the elbow flexed).

121 D

124 A

The Elbow
The bony landmarks were demonstrated in the previous section.

Observe flexion and extension at the elbow, and identify the muscles involved.

Demonstrate the action of biceps and triceps during supination. Why is the triceps active during supination?

130 A

139 C

The Wrist
Review the bony landmarks seen on the wrist in the previous section.
Identify the tendons on the flexor and extensor surfaces of the wrist.

Observe the movements at the wrist joint:
○ flexion,
○ extension,
○ abduction (radial deviation),
○ adduction (ulnar deviation),
and demonstrate the muscles active in these movements.

140 A,B,C

Demonstrate the movements and muscles involved in the fingers:
○ flexion,
○ extension,
○ adduction ⎱ to or from the bony axis of
○ abduction ⎰ the middle finger.

Can the middle finger be abducted and adducted (**1**)? What movements are permitted at the interphalangeal joints?

1

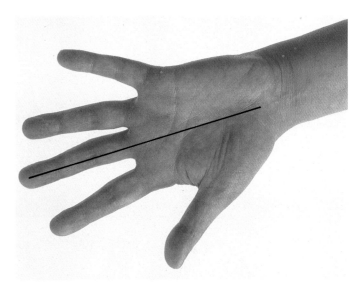

Movements of the thumb
These occur in a plane at right angles to the corresponding movements of the fingers:
○ flexion and extension of the thumb occur in the coronal plane (**2**),
○ abduction occurs forwards in the sagittal (anteroposterior) plane (**3**),
○ adduction occurs backwards in the saggital plane, which restores the thumb to its anatomical position after abduction (**3**),
○ opposition is an important movement peculiar to the thumb.

2

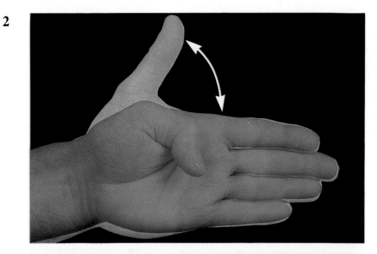

3

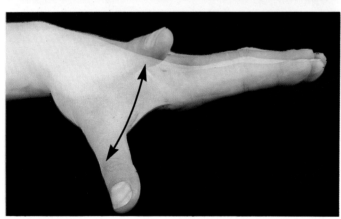

The thumb is pulled medially across the palm of the hand in such a way that the palmar surface of its tip meets the palmar surfaces of the fingertips (**4**).

4

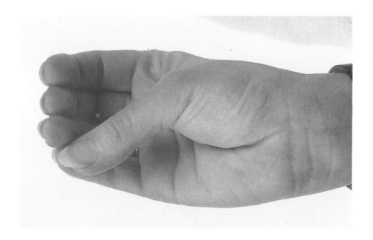

Position of rest (5)

Precision grip
This grip is for a delicate and precise control of an object, for example, a pen. The object is held between the thumb and fingers, the finest adjustments being carried out by the intrinsic muscles of the hand (**6**).

5

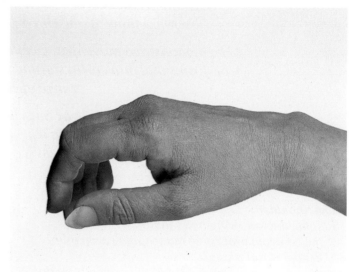

6

Power grip
The purpose of this grip is to hold an object firmly while it is moved by the action of the more powerful proximal muscles of the limb (**7**).

7

In each of the above cases, observe the state of flexion, extension and other movements at the wrist and metacarpophalangeal joints. Make a list of the muscles likely to be active, and note their nerve supply.

The anatomical 'snuff box' is an important topographical region. It is defined by the tendons of extensor pollicis longus and extensor pollicis brevis with abductor pollicis. In its floor there is the base of the first metacarpal, trapezium and scaphoid bone.

Superficial veins of the upper limb

Block the superficial viens by grasping the subject's arm, and ask the subject to repeatedly clench and release the fist. This will cause the superficial veins to stand out. Locate the **cephalic**, **basilic** and **median cubital veins**.

Arteries

Find the pulses of:

○ the **subclavian artery**, by pressing downwards behind the clavicle in the posterior triangle of the neck,

○ the **axillary** and **brachial arteries**, on the medial side of the humerus in the upper part of the arm,

○ the **brachial artery**, detected just medial to the bicipital tendon in the cubital fossa,

○ the **radial artery**, pressed against the radius lateral to the tendon of the flexor carpi radialis, and in the floor of the anatomical snuff box,

○ the **ulnar artery**, felt as it passes over the flexor retinaculum, just proximal to the hypothenar eminence,

○ the **superficial palmar arch**; this cannot be felt, but can be represented by drawing an arch on the palmar surface of the hand between the ulnar and radial arteries. The highest point on this arch lies on a line at the level of the distal border of the fully extended thumb.

Nerves

The **ulnar nerve** can be palpated posterior to the medial epicondyle of the humerus. If it is compressed on this point, its distribution in the hand can be located as a paraesthesia ('pins and needles') over the lateral part of the hand.

The **median nerve** can be palpated in the cubital fossa, medial to the brachial artery.

The **radial nerve** may sometimes be palpated in the lower part of the arm, between the brachialis and brachioradialis muscles. Its superficial branch crosses the anatomical snuff box, and may be rolled against the tendon of extensor pollicis longus.

Synovial sheaths of the hand (8)

The **common synovial sheath** for the tendons of **flexor digitorum superficialis** and **flexor digitorum profundus** extends proximally to the flexor retinaculum, to almost as far as the distal palmar skin crease, with a prolongation into the terminal phalanx of the little finger.

The **synovial sheath for flexor pollicis longus** extends proximally to the flexor retinaculum, to the terminal phalanx of the thumb.

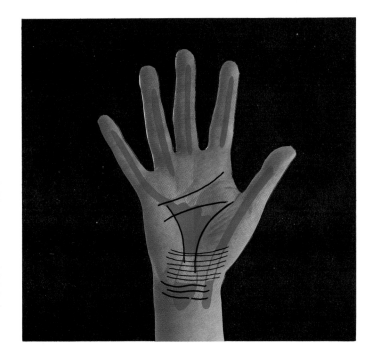

8

Thorax

37

The Thoracic Wall

The purpose of this dissection is to display the muscles covering the thoracic wall, and to demonstrate the muscles, nerves and vessels of the intercostal spaces.

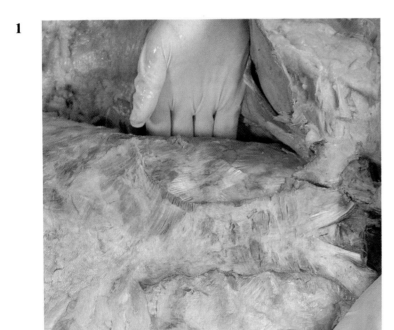

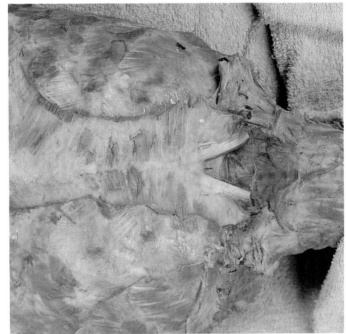

If the upper limb has not been dissected, start with the breast and pectoral region (sections 23 and 24).

*Revise the attachments of the **pectoralis major** and **pectoralis minor** muscles to the thoracic wall.*

*Identify the **serratus anterior muscle**.*

- Detach the slips of the serratus anterior from the ribs. Pass your hand between the serratus anterior and the thoracic wall, and separate the muscle from the thoracic wall (**1**).

- Clean the thoracic wall, leaving a small tag of each of the upper limb muscles to remind you of their origins (**2**).

155 B

*Observe the **external intercostal muscles** and **anterior intercostal membranes**. The **internal intercostal muscles** can be seen through the anterior intercostal membranes.*

Note the directions of the fibres of both sets of intercostal muscles.

- Dissect away the external and internal intercostal muscle from one intercostal space (**3**).

● Starting as far posteriorly as is possible, use rongeurs to nibble away the lower border of the rib, taking care to avoid damaging the structures lying deep to it .

Repeat the dissection of the intercostal nerve in one of the lower spaces, in order to demonstrate that the lower nerves leave the intercostal spaces to innervate the anterior abdominal wall.

156 A *Identify the intercostal nerve and vessels lying on the innermost intercostal muscle.*

Note the relations of the structures to each other (4).

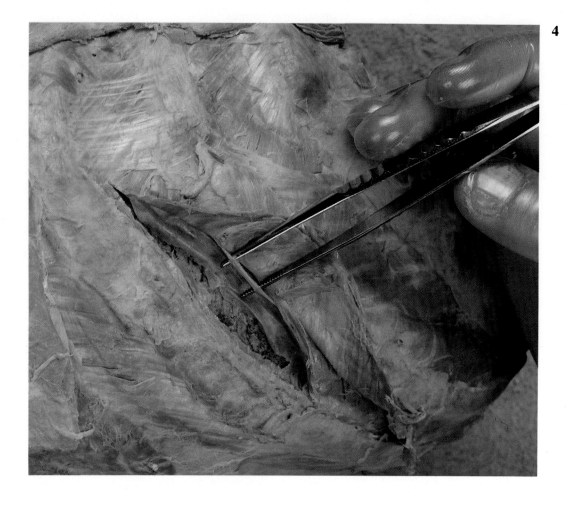

4

38

The Lungs and Pleura

In this dissection the anterior wall of the thorax is removed and the thoracic contents are studied in situ. Then the lungs are removed and studied, and the impressions on their medial surfaces are matched with the structures in the mediastinum. Finally, the pleural cavity is explored.

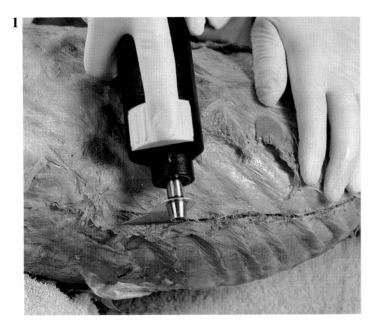

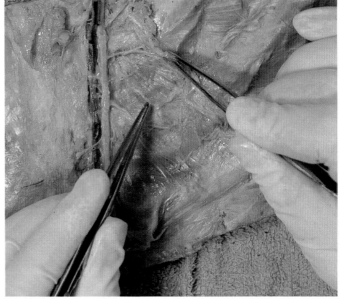

Use a saw to cut through the manubrium of the sternum, just below its joints with the first costal cartilages.

Make a second saw cut through the xiphisternal joint and continue this cut posteriorly above and parallel to the costal margin, and through the seventh and eighth ribs. Avoid damage to the diaphragm by pushing the soft tissues inwards as you use the saw.

Divide the ribs in the posterior axillary line, or further posteriorly if possible (**1**).

Lever off the upper part of the thoracic wall. Divide the internal thoracic vessels and the fibrous tissue which binds the pericardium to the posterior surface of the sternum.

Finally, divide the internal thoracic vessels or their terminal branches at the level of the xiphisternal joint and lift off the anterior wall of the thorax.

58 A,B   *Examine the posterior surface of the anterior thoracic wall.*

*Observe the extent of the **parietal pleura**, especially near the midline.*

*Strip off the parietal pleura to observe the internal thoracic vessels and their terminal branches (**2**).*

Observe the innermost intercostal muscles.

Retain this part of the thoracic wall carefully, and replace it from time to time during the dissection to check the position of the deeper structures with respect to the ribs.

*Observe the **pericardium** and **lungs** in situ (**3**).*

Define the relations of the pleura to the front of the pericardium.

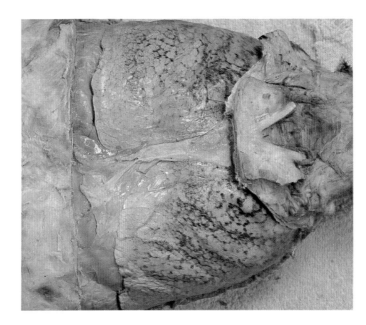

3

159

Pass your hand around the posterior surface of the lung to break down any pleural adhesions.

178-
183

*Locate the **oblique fissures** of both lungs, and the **horizontal fissure** of the right lung.*

*Define the **upper**, **middle** and **lower lobes** of the lungs.*

Replace the anterior thoracic wall and check the levels of the fissures with respect to the ribs.

*Pull one of the lungs laterally, to expose the pleura covering the pericardium (**mediastinal pleura**). Observe the **lung root** and palpate the structures within it.*

4

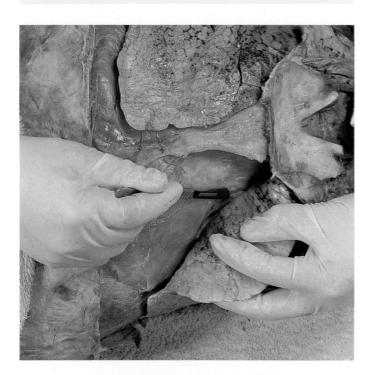

5

Pull the lung further laterally, and divide the lung root as close to the lung as possible. Remove the lung (**4**). In a similar way, remove the lung on the opposite side.

Study the lungs carefully, observing their surfaces and borders:
○ *the **base**, which is concave and fits over the dome of the diaphragm,*
○ *the **sternocostal surface** fits against the ribs and sternum.*
○ *the **posterior surface** fits in the paravertebral gutter,*
○ *the **medial surface** is in contact with the vertebral column and mediastinum. The mediastinal part of the medial surface carries the impressions of several structures which project from the mediastinum;*
○ *the **apex**.*

Explore the pleural cavities with your fingers. Find: **190 C**
○ *the **paravertebral gutters**,*
○ *the **domes of the pleura**,*
○ *the **costodiaphragmatic recesses** (**5**).*

Identify the structures which pass into the lung through the lung root:
○ *the **bronchi**,*
○ *the **branches of pulmonary arteries** and **veins**.*
*Identify the **pulmonary lymph nodes**.*

Use forceps to remove some of the lung tissue at the root of the lung in order to study the relations of the bronchi and pulmonary vessels more closely.

Follow the branches of the bronchus (*lobar bronchi*) into the lobes of the lung, and trace one or two of them   **176** with their accompanying arteries, into the lung substance as far as the *segmental bronchi*, which supply the *bronchopulmonary segments*.

Now study the lateral walls of the mediastinum; you   **172**
may have to feel for some of these structures rather   **173**
than see them. Match what you find to the medial   **174**
surface of the lung.   **175**

On the left side locate (6):
○ the **aorta**,
○ the **common carotid artery**,
○ the **subclavian artery**,
○ the **left ventricle**,

○ the **infundibulum**,
○ the **oesophagus**.
Match these with their impressions on the medial surface of the left lung.

6a

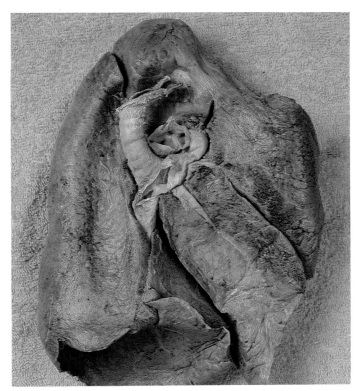

6b

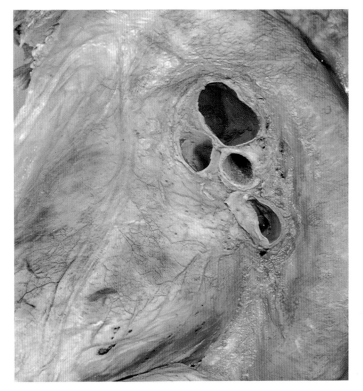

7a

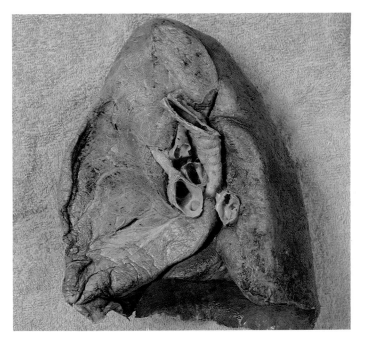

7b

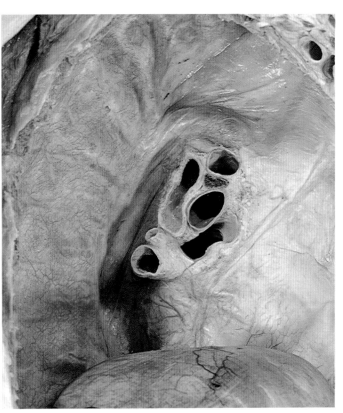

On the right side locate (7):
○ the **oesophagus**,
○ the **trachea**,
○ the **superior vena cava**,
○ the **azygos vein**,
○ the **inferior vena cava**,
○ the **right atrium**.
Match the structures with their impressions on the medial surface of the right lung.

Observe the relationship of the impressions on the lungs to the lung roots. Match the structures on the cut surface of the lung roots to the same structures in the mediastinum.

186 A,B

39

The Middle Mediastinum and Heart

In this dissection the pericardium, the exterior of the heart and blood supply of the heart are studied.

1

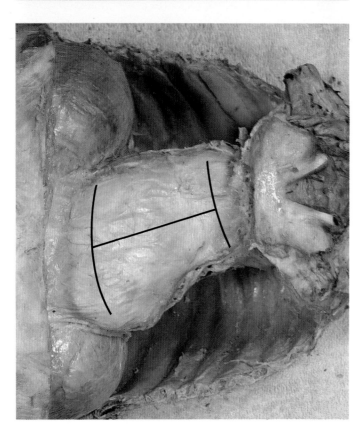

Review all visible and palpable structures which were observed in the mediastinum during the dissection of lungs and pleura.

Remove the parietal peritoneum piecemeal from the accessible part of the mediastinum (**1**).

*Identify the **right and left phrenic nerves** and the **pericardiophrenic vessels** which accompany them.*

Confirm that these structures pass in front of the lung root, and that the right phrenic nerve is related to the superior vena cava, the right atrium and the inferior vena cava.

Make the incisions in the wall of the pericardium as shown in (**2**). This is done by pinching up a fold of the pericardial wall and cutting through it with scissors. Take care not to damage the heart.

Turn back the flaps of the pericardial wall to expose the heart (**3**).

2

3

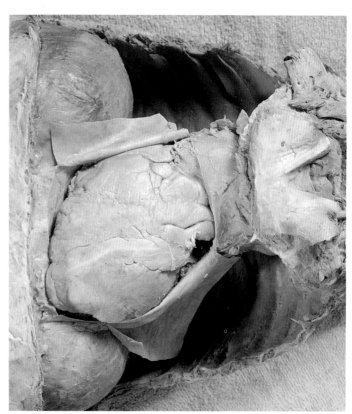

Observe the three layers of the pericardium:
- *the outermost **fibrous layer**,*
- *the **parietal serous layer** which lines the fibrous layer,*
- *the **visceral serous layer** which invests the heart.*

*Explore the pericardial cavity with your fingers. Find the roots of the **aorta, superior** and **inferior venae cavae** and the **pulmonary vessels** within the pericardium.*

Push your finger upwards under the uppermost horizontal incision, to establish the position of the reflection of the pericardium on to the great vessels.

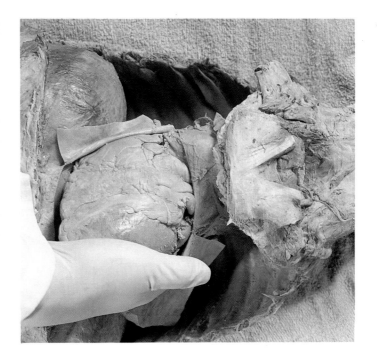

160 C

*Lift up the apex of the heart and push a finger into the **oblique sinus of the pericardium** (**4**). Observe that a finger pushed into the pericardium to the right of the heart cannot enter the oblique sinus.*

160 B,C

*Grasp the ascending aorta and pulmonary trunk between your thumb and forefinger. The tips of your thumb and forefinger meet in the **transverse sinus of the pericardium** (**5**).*

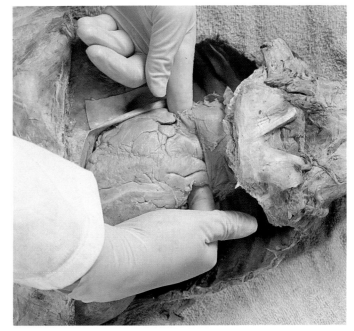

160 A

With the heart in situ *define:*
- *the **anterior** or **sternocostal surface**,*
- *the **base** and **apex**,*
- *the **inferior** or **diaphragmatic surface** and*
- *the right and left borders of the heart.*

Observe that:
- *the sternocostal surface is formed by the wall of the right ventricle,*
- *the right border is formed by the right atrium,*
- *the left border is formed by the left ventricle and the auricle of the left atrium.*

To remove the heart:
- Place a probe in the transverse sinus of the pericardium and cut down through the aorta and pulmonary trunk to the probe (**6**).

- Cut through the superior vena cava below the entrance of the azygos vein.

- Cut through the inferior vena cava as close as possible to the diaphragm.

- Divide the two layers of serous pericardium between the inferior vena cava and the right pulmonary veins (this reflection of serous pericardium forms the right wall of the oblique sinus of the pericardium).

- Divide the layers of serous pericardium between the left and right pulmonary veins (this reflection contributes to the partition between the oblique and transverse sinuses).

The heart can now be lifted out of the pericardium. There is often a considerable amount of fat on the heart beneath the serous pericardium. Remove this fat carefully whenever it gets in the way.

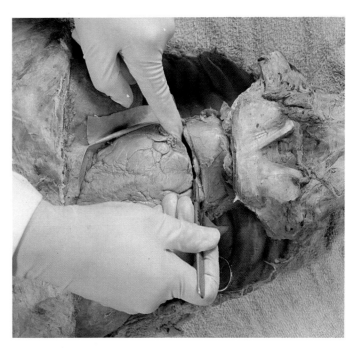

Find the borders of the four chambers of the heart.

*The **atrioventricular grooves** separate the atria from the ventricles and contain the **coronary vessels**.*

*The **anterior** and **inferior interventricular grooves** denote the position of the **interventricular septum**.*

*On the posterior aspect or base of the heart the four pulmonary veins run into the **left atrium**.*

*Find the **left** and **right auricles**.*

Locate the coronary arteries in·the atrioventricular (coronary) groove. Clean the arteries and surfaces of the ventricles to demonstrate the *anterior interventricular* and circumflex branches of the left coronary artery, and the *marginal and posterior interventricular branches* of the right coronary artery. **161 D,E**

Locate the coronary sinus in the atrioventricular groove. Trace its larger tributaries back to the walls of the ventricles (**7**). Variations in the pattern of arterial supply and venous drainage of the heart are common. Compare the arteries and veins of your specimen with the examples shown in the atlas and in standard textbooks of anatomy. **165-171**

7

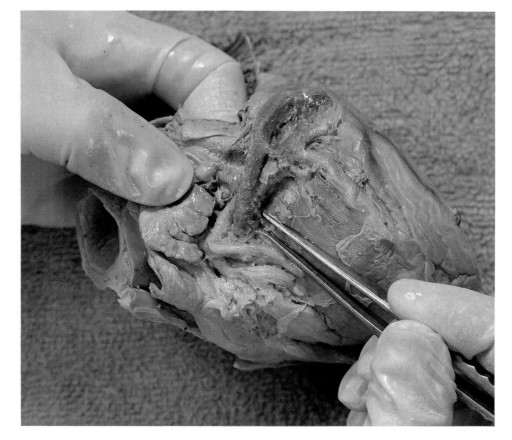

40

The Interior of the Heart

*In this dissection the chambers of the heart are opened in turn and
explored both visually and with your fingers*

Open the right atrium by making a short incision from
the inferior vena cava to the base of the right auricle,
then continue this incision posteriorly (**1**). Remove the
blood clot and wash out the atrium with cold water.

162 A

Identify:
- ○ *the smooth **atrial wall**,*
- ○ *the **musculi pectinati** and the **crista terminalis**,*
- ○ *the **orifices of the superior and inferior venae cavae
  and the coronary sinus**,*
- ○ *the **fossa ovalis**,*
- ○ *the **right atrioventricular orifice**.*

Open the right ventricle by removing most of its
anterior wall (**2**). Open the pulmonary trunk. Remove
the blood clot; wash out the ventricle in cold water.

162 B

Now identify the following structures:
- ○ *the **tricuspid valve, chordae tendineae** and **papillary
  muscles**,*
- ○ *the **trabeculae carnae** and the **septomarginal
  trabecula**,*
- ○ *the **infundibulum**,*
- ○ *the **pulmonary orifice and valve**.*

*The atrioventricular valve can be seen clearly when a
finger is passed through it from the right atrium (**3**).*

164 A

*Trace the course of blood through the right ventricle
from the atrioventricular valve to the pulmonary
orifice.*

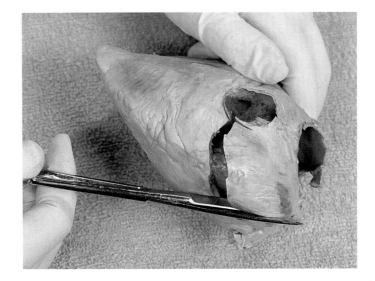

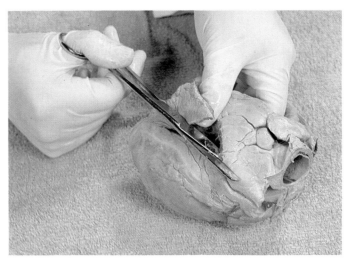

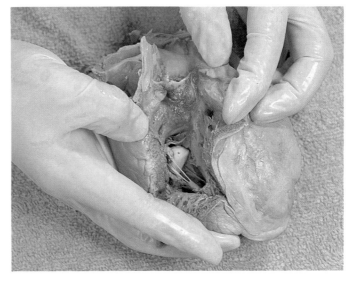

Open the left atrium with an inverted U-shaped incision in its posterior wall. Remove the blood clot and wash it out as before (4).

4
164 B

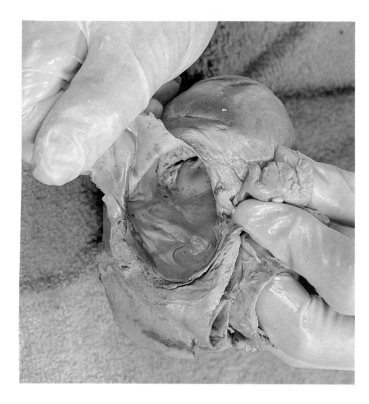

Open the left ventricle by making a cut from the apex of the heart about one centimetre from (and parallel to) the interventricular groove, to the root of the ascending aorta (5).

5
163 C

Open the ascending aorta with an incision between the left and right coronary arteries.

164 B,
C

41

The Superior and Posterior Mediastinum

*In this dissection the structures in the superior and posterior mediastinum
and the posterior wall of the thorax will be studied.*

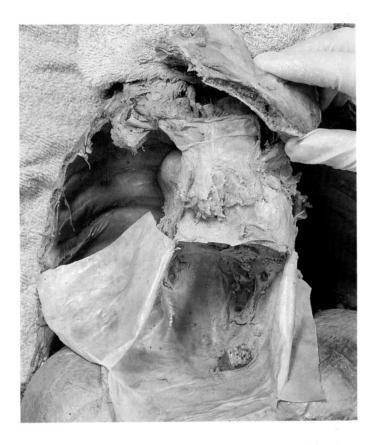

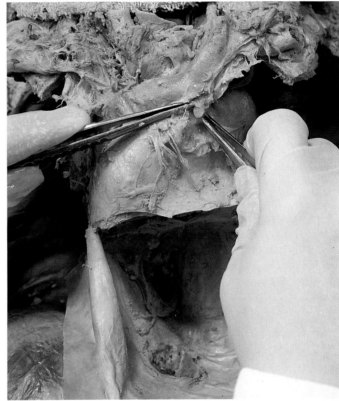

The superior mediastinum is located in front of the first four thoracic vertebrae, between the superior aperture of the thorax and sternal angle.

- If the neck has been dissected, saw through the first costal cartilages to free the top part of the manubrium of the sternum (1).

- If the neck has not been dissected, leave the top part of the sternum *in situ* and postpone its dissection until after the dissection of the neck.

Carefully remove the loose connective tissue between the structures in the superior mediastinum (2).

188 B
189 C
*Find the tributaries of the superior vena cava, the two **brachiocephalic veins** and the **azygos vein**.*

*Find the branches of the arch of the aorta, the **brachiocephalic, left common carotid** and **subclavian arteries**.*

Clean the arch of the aorta, pulmonary trunk and pulmonary arteries. Clean the lower part of the trachea and follow the main bronchi and pulmonary arteries towards the site of the lung root.

*Locate the **ligamentum arteriosum** between the arch of the aorta and the left pulmonary artery.*

Note the relations of the great vessels to each other and to the trachea and bronchi.

Follow the vagus and phrenic nerves in the superior mediastinum.

Replace the head and neck specimen and trace structures from the neck through the thoracic inlet into the superior mediastinum.

190 B

*Place your fingers in the **dome of the pleura**; note its relations in the neck, the **subclavian vessels** and their branches, and the **brachial plexus**.*

The posterior mediastinum lies behind and below the pericardium, in front of the bodies of the fifth to the twelfth thoracic vertebrae.

Remove the posterior wall of the pericardium, but be careful to preserve both phrenic nerves and their accompanying vessels **(3)**.

Locate the oesophagus in the loose connective tissue of the posterior mediastinum, and clean it.

*Find the **right** and **left vagus nerves** on the upper part of the oesophagus, and note that they form the oesophageal plexus on its lower part. Note that the oesophagus inclines to the left in the lower part of the posterior mediastinum.*

194 A

*Move the oesophagus over to the left, to see the **azygos vein**. Note the tributaries which come from the right intercostal spaces. Move the oesophagus over to the right and examine the descending aorta.*

*Use forceps or a blunt probe to search between the aorta and the azygos vein for the delicate **thoracic duct**. When you have found it, look for the **hemiazygos veins** on the left side and their communications with the azygos vein.*

Replace the heart and take note of its posterior relations.

The following structures are on the posterior wall of the thorax. Peel the parietal pleura carefully away from the bodies of the vertebrae and the floor of the paravertebral gutter.

194 B

*Find the posterior **intercostal vein, artery and nerve** in one intercostal space. Try to demonstrate the communications between the nerve and its sympathetic ganglion.*

*The **sympathetic trunks** lie on the necks of the ribs. Trace the **greater splachnic nerve** (T5–T9) and the **lesser splachnic nerve** (T10–T11) into the posterior mediastinum.*

Finally, using the head and neck specimen and this dissection, follow the vagus nerve from the cranial cavity through the neck and thorax, and the phrenic nerve from the neck to the diaphragm, taking note of their relations in the neck and thorax.

3

42

Osteology and surface anatomy of the Thoracic Cage

The individual bones are studied first, then the thoracic cage is studied in an articulated skeleton. Finally, the thoracic cage is outlined on a living subject.

Isolated bones

7-151   Refer to the atlas for the detailed anatomy of the thoracic vertebrae, ribs and sternum.

Articulation of a rib with the vertebral column

147 C   Articulate a typical rib with two typical thoracic vertebrae. The lower facet on the head of the rib articulates with the half facet on the superolateral aspect of the body of the corresponding vertebra. The upper facet on the head of the rib articulates with a similar half facet on the inferolateral aspect of the vertebra above. The facet on the tubercle of the rib articulates with a facet on the transverse process of the corresponding vertebra, that is, the eighth rib and eighth transverse process.

Articulated skeleton
The thoracic part of the vertebral column. Note:
● The curvatures of the vertebral column.

● The intervertebral discs.

● The articulations of the typical ribs with the adjacent vertebrae and intervertebral disc.

● Which ribs do not articulate in this typical manner?

● The intervertebral foramina formed from adjacent vertebral notches.

● The spinous processes. Note that their direction and size change from above downwards. What is the difference in level between the tip of the spinous process and the body of the vertebra to which it belongs?

● The joints between the articular processes of neighbouring vertebrae. What movements can occur at these joints?

Thoracic cage (**1**)
● Take note of the way the ribs are arranged in the articulated skeleton. They pass obliquely downwards and forwards from the vertebral column.

● The costal cartilages of a typical rib slope upwards to the sternum.

● The intercostal spaces are wider in front than behind.

● Distinguish between true ribs, false ribs and floating ribs.

● Identify the thoracic inlet and bones which form its borders.

● Note the increase in transverse diameter of the thoracic cage. Which rib is at the maximum diameter of the thoracic cage?

● Identify the thoracic outlet and skeletal structures which form its borders. Take particular care to appreciate the way in which the lower costal cartilages articulate with each other, and describe the formation of the costal margin.

1

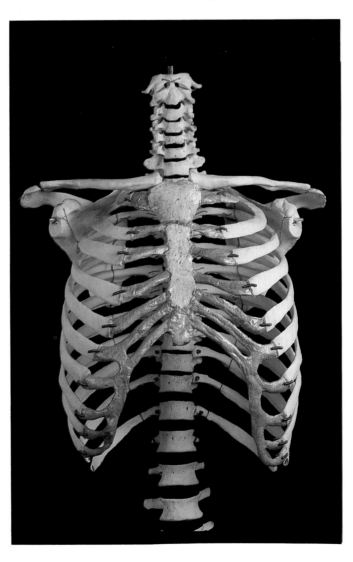

Surface anatomy of the thoracic skeleton (2)

Find the following structures on a living subject by palpation. Some of them may be visible, but mark all of them with a skin pencil:

- clavicle,
- manubrium, body and xiphoid process of the sternum
- jugular notch,
- sternoclavicular joint,
- sternal angle,
- costal margin,
- spinous process of the seventh cervical vertebra.

Show how to identify ribs starting with the second rib at the manubriosternal joint.

Answer the following questions, by referring to the living subject and articulated skeleton:

- Which costal cartilage articulates with the sternum at the sternal angle?

- Can the first rib be palpated in the living subject?

- Which costal cartilage forms the lowest part of the costal margin?

- Can the eleventh and twelfth ribs be palpated?

- What is the position of the nipple in the male?

- How many ribs can be identified in the axilla?

- Can the spines of all the thoracic vertebrae be palpated?

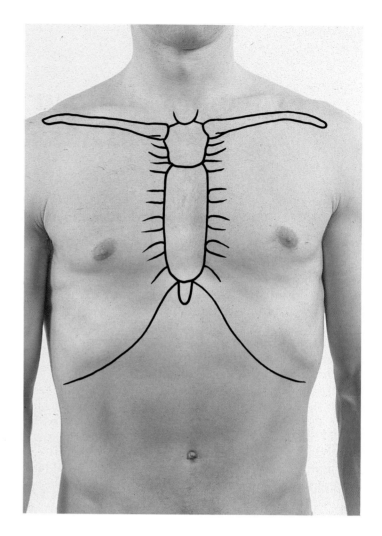

43

Surface Anatomy of the Pleura, Diaphragm and Lungs

You should be able to identify the bony landmarks of the thoracic wall in the living subject. You will also need to refer to the mid-clavicular line, the mid-axillary line, and to a line corresponding to the tips of the transverse processes of the thoracic vertebrae, about one inch or three centimetres from the midline.

Pleura
Map out the surface projections of the pleura on an articulated skeleton. Then, using a skin pencil, outline those onto a living subject.

The anterior surface markings of the pleura (1)
On the right side:

- Draw a line from a point two to three centimetres above the medial third of the clavicle to a point just to the right of the midline at the sternal angle. The anterior border then runs straight down to the sixth costal cartilage.

On the left side:

- The anterior border of the pleura is parallel to that on the right side, as far as the fourth costal cartilage; then it is deflected laterally and runs down to the sixth costal cartilage, along the lateral margin of the sternum.

The inferior surface markings of the pleura
The reflection of the pleura crosses the eighth rib at the mid-clavicular line, the tenth rib at the mid-axillary line, and the twelfth rib at a line corresponding to the transverse processes of the thoracic vertebrae.

The posterior surface markings of the pleura (2)
This runs straight up from the point where the line crosses the twelfth rib to the level of the spine of the first thoracic vertebra.

Auscultation
Use the stethoscope to listen to normal breathing sounds laterally in the third intercostal space.

Percussion
Place one finger on the chest wall and tap it with the index finger of the other hand. Get one of the demonstrators to show you how to do this properly.

1
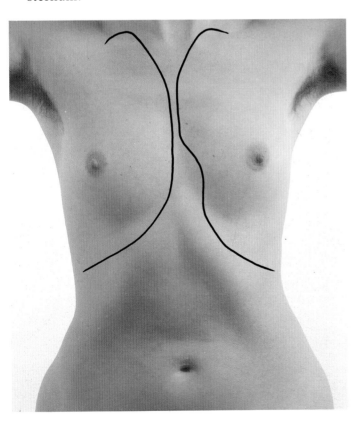

2
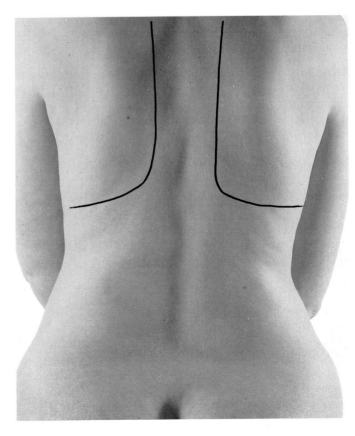

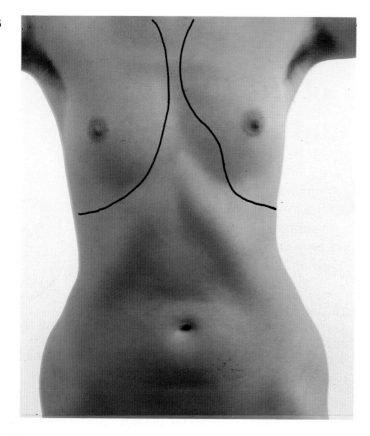

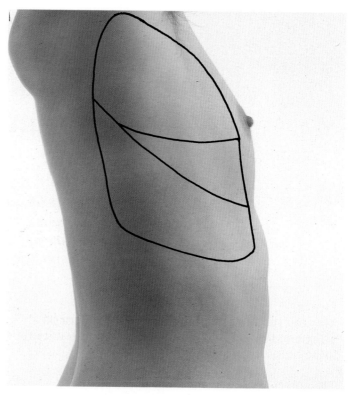

Solid organs produce a dull note, hollow organs produce a more resonant note. Use percussion to determine the extent of the lungs, heart and liver in your subject. (See how the areas defined by percussion coincide with your surface markings as indicated earlier in the section.)

Movements of respiration
Feel the increase in the anteroposterior dimensions of the thorax during inspiration, by placing one hand on the sternum and the other on the back of your subject, at the same level.

Feel for the increase in the transverse diameter with your hands on the mid-axillary lines.

Observe the anterior abdominal wall in inspiration and expiration.

Use a tape measure to measure the circumference of the chest at the level of the fourth costal cartilage:
● in quiet inspiration,
● in quiet expiration,
● in deep inspiration,
● in full expiration.
Measure also the change in circumference of the abdomen at the level of the umbilicus.
What does this reveal about the role of the abdominal muscles in respiration?

Lungs (3)
The anterior border of the right lung is just lateral to the border of the right pleural sac, as far as the sixth costal cartilage. The anterior border of the left lung follows the pleura as far as the fourth costal cartilage. The border is then represented by a line drawn from the fourth chondrosternal joint to a point reaching one inch to the left of the sternum on the sixth costal cartilage.

The inferior borders of the lungs cross:
● the sixth rib at the mid-clavicular line,
● the eighth rib at the mid-axillary line,
● the tenth rib at the paravertebral lines.
The posterior borders follow the posterior borders of the pleura.

Fissures of the lungs (4)
● Draw a line from the spinous process of the third thoracic vertebra to the sixth costal cartilage, five centimetres from the midline. This will give you the position of the oblique fissure, which is the same on both sides.

● For the transverse fissure (on the right side only), draw a horizontal line from the fourth costal cartilage to meet the oblique fissure.

Diaphragm (5)

Draw in the level of the domes of the diaphragm on the living subject, during:

- **Quiet inspiration:** Right: fifth intercostal space.
  Left: sixth rib.

- **Quiet expiration:** Right
  Left: 1.5 centimetres higher than in quiet inspiration.

- **Deep inspiration:** Right: sixth intercostal space.
  Left: seventh rib.

- **Deep expiration:** Seven centimetres higher than in deep inspiration, that is, at the level of the third rib.

Thus, several of the abdominal organs, although below the diaphragm, are within the thoracic cage. These include the spleen, most of the liver, most of the stomach and the upper parts of the kidneys.

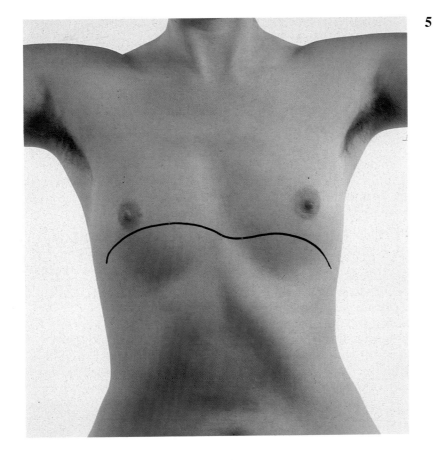

5

44

Surface Anatomy of the Heart and Mediastinum

It is most convenient to sketch the surface markings of the heart on a male subject, but every student should locate the apex beat and surface markings of the mediastinal structures in both sexes.

Heart and great vessels on the living subject (1)
Find the sternal angle, xiphisternal joint and fifth left intercostal space:
- Draw a curved line from the midpoint of the sternal angle to a point four centimetres to the right of the midline at the xiphisternal joint.

- Draw a second curved line from the midpoint of the sternal angle to a point in the fifth left intercostal space, eight centimetres from the midline at the xiphisternal joint.

- Connect the lower ends of these two lines with a straight line which slopes slightly upwards to the left.

Refer to your dissected specimen to indicate the right and left ventricles; refer to your outline to indicate the right atrium and the two atrial appendages. These markings are approximate. The size and shape of the heart varies according to the physique of the individual. Its position also depends on respiratory movements (since the heart moves with the diaphragm), and on whether the subject is standing up or lying down.

Find the apex beat. Ask the subject to sit leaning forwards slightly. The apex beat should be visible in a thin subject; it can be felt even when it cannot be seen. It is usually in the fifth left intercostal space, eight centimetres from the midline.
Compare this position with the position of the apex beat when the subject lies on his or her back.

Auscultation
Use a stethoscope to listen to the sounds of the heart at the apex ('lub-dupp'). The first sound results from the closure of the two atrioventricular valves, and the second from the closure of the pulmonary and aortic valves.

Heart valves
Plot the positions of the heart valves:
- The pulmonary valve lies behind the sternal end of the third left costal cartilage.

- The aortic valve lies behind the left half of the sternum, opposite the third intercostal space.

- The mitral valve lies directly below the aortic valve, opposite the fourth costal cartilage.

- The tricuspid valve lies behind the right half of the sternum, opposite the fifth costal cartilage.

Indicate other mediastinal structures (1):
The line drawn from the sternoclavicular joint, along the right margin of the heart, represents:
- the right brachiocephalic vein,
- the superior vena cava,
- the right atrium,
- the inferior vena cava.

The left brachiocephalic vein crosses behind the upper half of the manubrium of the sternum, from the left sternoclavicular joint.

The arch of the aorta winds posteriorly behind the lower half of the manubrium of the sternum. Draw in the ascending aorta, from the valves to the aortic arch.

The bifurcation of the trachea is located at the level of the sternal angle.

When the surface markings of these structures have been indicated on the living subject, check the results against their positions in the dissected specimen.

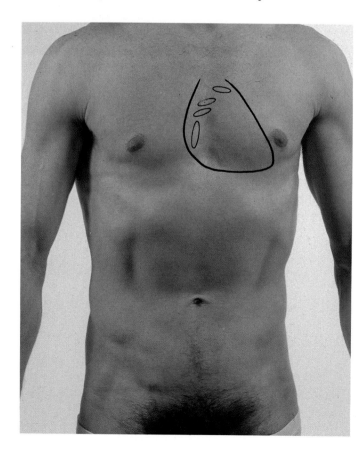

1

Abdomen

45

The Anterior Abdominal Wall

The lateral part of the anterior abdominal wall consists of eight layers: skin, two layers of superficial fascia, three layers of muscle, a further layer of fascia, and the peritoneum. The medial part, deep to the superficial fascia, consists of the rectus abdominis muscle in its sheath. The purpose of this dissection is to demonstrate these layers of the abdominal wall and·to show how they contribute to the rectus sheath.

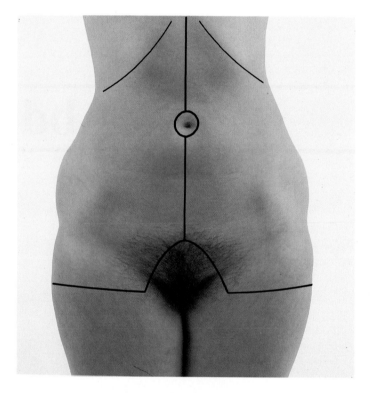

Before starting to dissect, find the following bony landmarks on the cadaver so that you understand the extent of the anterior abdominal wall:
- *the costal margin,*
- *the pubic symphisis and tubercle,*
- *the anterior superior iliac spine,*
- *the iliac crest.*

Find also the linea alba, linea semilunaris and inguinal groove.

Make the incisions shown in the illustration (1):
- *a vertical incision from the xiphisternum to a point an inch or so above the pubic symphysis,*

- *a horizontal incision, on each side, from the lateral side of the thigh towards the midline. Connect the two horizontal incisions with a further incision encircling the external genitalia. Leave the skin on the external genitalia in both sexes to prevent drying out.*

Reflect the skin flaps laterally, preserve them to protect the specimen in the later stages of the dissection. It may be helpful to make a second horizontal incision on both sides which passes through the highest point of each iliac crest.

Find the anterior terminal branches and the lateral branches of the intercostal nerves and vessels in the superficial fascia. Identify and preserve them.

Cut through the superficial fascia along a line between the highest points of the two iliac crests.

Separate with your fingers the superficial fascia from the muscle of the abdominal wall.

Demonstrate the existence of the fatty layer (of Camper) and deeper membranous layer (of Scarpa) (2)

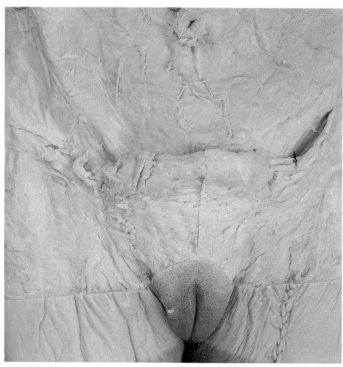

*Push your fingers deep to the membranous layer into the thigh; you will meet resistance where the membranous layer fuses with the deep fascia of the thigh, the **fascia lata**.*

*Using your fingers, define the extent of space between the membranous layer and muscles of the abdominal wall (**3**)*

Demonstrate the communication between this space and the spaces within the penis and scrotum.

Remove the rest of the superficial fascia except for a small tag to show where this attaches to the fascia lata.

Clean the surface of the external oblique muscle and its aponeurosis. Avoid damaging the superficial inguinal ring.

204 A *Note the full extent of the **external oblique muscle** and the direction taken by its fibres.*

Note the extent of its fleshy and aponeurotic parts.

206 A *Note its posterior free edge and its interdigitation with the origin of the serratus anterior muscle.*

Open the rectus sheath by means of a longitudinal incision medial to the anterior cutaneous nerves and vessels (**4**). Reflect the sheath away from the muscle, taking care as you separate it from the muscle at the tendinous insertions.

204 A
205 B Divide the rectus abdominis muscle close to its inferior
207 B attachment, and reflect it superiorly.

Study the posterior wall of the rectus sheath. Find:
  ○ *the **pyramidalis muscle**,*
  ○ *the **superior and inferior epigastric vessels**,*
  ○ *the **arcuate line**,*
  ○ *the **transversalis fascia** below the arcuate line (**5**).*

3

4

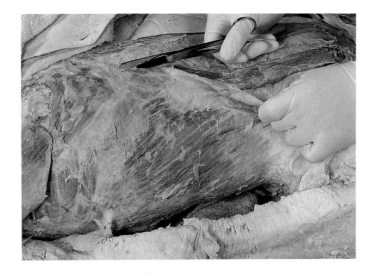

Make a vertical incision in the external oblique muscle, in the mid-axillary line. Make a second, horizontal incision from the highest point on the iliac crest to the linea semilunaris. Detach the muscle from its costal origins and reflect the upper part medially, leaving the posterior and inferior parts *in situ* (**6**).

Find the plane between the external and internal oblique muscles and separate them with your fingers. Clean the surface of the internal oblique muscle.

Note the extent and the direction taken by the fibres of **205 C**
*the **internal oblique muscle**.* **207 C**

5

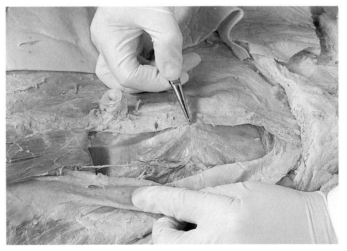

6

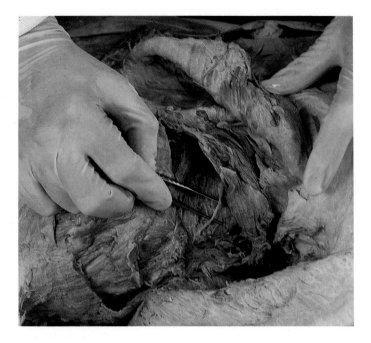

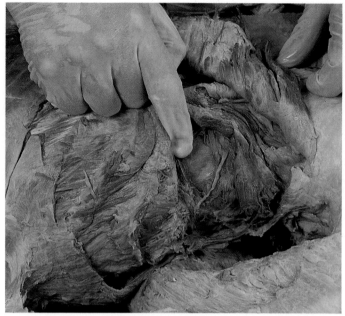

Detach the internal oblique muscle from its costal origin, and cut through it in the mid-axillary line and along the horizontal section in the external oblique muscle.

Find the plane between the internal oblique muscle and the transversus abdominis muscle, separate these muscles with your fingers and reflect the internal oblique muscle medially as far as the lateral border of the rectus sheath.

Clean the surface of the transversus abdominis muscle, but preserve the nerves and vessels lying on it.

*Note the direction taken by the fibres of the **transversus abdominis muscle**, and identify the nerves and vessels on its superficial surface (7).*

*There is no need to reflect the transversus abdominis muscle; just push the fibres apart and find deep to it the thin **transversalis fascia**, **extraperitoneal fat** and **parietal peritoneal membrane** (8).*

Review the layers of the abdominal wall which you have demonstrated:
- *the peritoneum,*
- *the transversalis fascia,*
- *the transversus abdominis muscle,*
- *the internal oblique muscle,*
- *the external oblique muscle,*
- *the membranous layer of the superficial fascia,*
- *the fatty layer of the superficial fascia,*
- *the skin.*

46

The Inguinal Canal and Scrotum

The purpose of this dissection is to help you understand the structure of the spermatic cord and coats of the testis, and the structure of the inguinal region. Although homologues of certain structures can be found in both sexes, there is little point in trying to dissect this region in the female.

Review all the layers demonstrated in the dissection of the anterior abdominal wall.

Read a simple account of the origin and descent of the testis.

Use your fingers to separate the layers of the body wall below the horizontal incisions of the previous dissection.

Make an incision around the root of the penis which continues vertically through the skin of the scrotum on both sides. Reflect the skin flaps laterally (**1**).

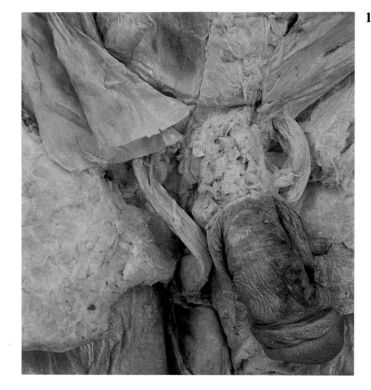

*Observe the **dartos muscle** lining the scrotum, and demonstrate the continuity of the membraneous layer of superficial fascia with*
○ *the dartos,*
○ *the superficial fascia of perineum and root of the penis.*
The dartos muscle forms a septum which divides the scrotum into two compartments.
*Observe the **fundiform ligament of the penis**, note its attachments to the body wall.*

*Find the **superficial inguinal ring** and note its shape, the medial and lateral crura and intercrural fibres. Observe that the **external spermatic fascia** is attached to the margins of the superficial inguinal ring (**2**).*

*Define the position of the inguinal ligament. **Gently separate** the scrotal wall from the spermatic cord and follow the cord into the scrotum.*

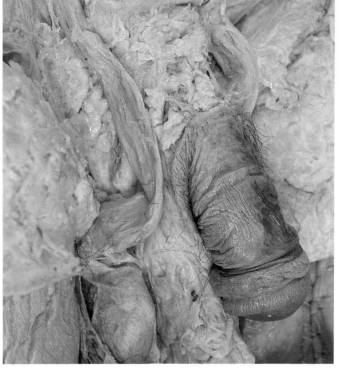

Make a vertical incision in the lower part of the external oblique aponeurosis just lateral to the rectus sheath which ends just medial to the superficial inguinal ring.

*Look at the inner surface of the **external oblique muscle**, the upper surface of the **inguinal ligament**, and find the **lacunar ligament**.*
*Find the **ilioinguinal** and **iliohypogastric nerves** lying on the surface of the internal oblique muscle. These nerves pierce this muscle close to the anterior superior iliac spine.*

*Note the appearance of the **internal oblique muscle**
and its division into muscular and tendinous parts.
Muscle fibres which have their origins on the lateral
part of the inguinal ligament run medially towards the
superficial inguinal ring.*

*Note the insertion of the internal oblique muscle into
the **falx inguinalis** (conjoint tendon).*

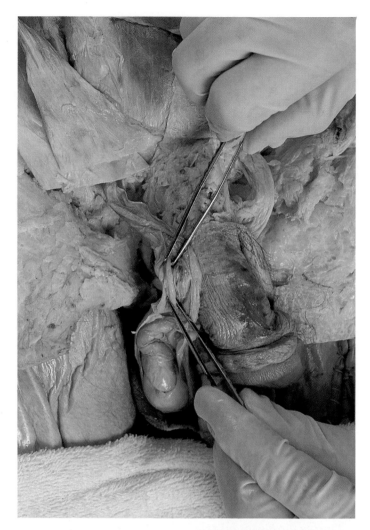

Incise the outermost layer of the covering of the
spermatic cord and testis, the *external spermatic fascia*.
Do not open the superficial inguinal ring. This will
hold the dissection together for you.

*Observe the **cremaster muscle** in the spermatic cord
and looping round the testis.*

Divide the cremaster muscle and fascia over the testis.

*Observe the peritoneal investment of the testis, the
tunica vaginalis testis.*

Incise the parietal layer of the tunica vaginalis testis on
one side.

Explore the space between the testis and its coverings.

*Observe the testis and the **epididymis**; note the **sinus of
the epididymis**.*

*Pay particular attention to the orientation of the testis
within the scrotum.*

Relate the various layers of:
○ *the coverings of the testis,*
○ *the layers of the spermatic cord (**3**),*
○ *the layers of the anterior abdominal wall to each
other.*

*Demonstrate the **inguinal canal** (**4**):*
○ *its **floor**, the inguinal ligament,*
○ *its **posterior wall**, the conjoint tendon and
transversalis fascia,*
○ *its **anterior wall**, the aponeurosis of the external
oblique muscle,*
○ *its **roof**, the internal oblique muscle.*
Identify again the deep and superficial rings.

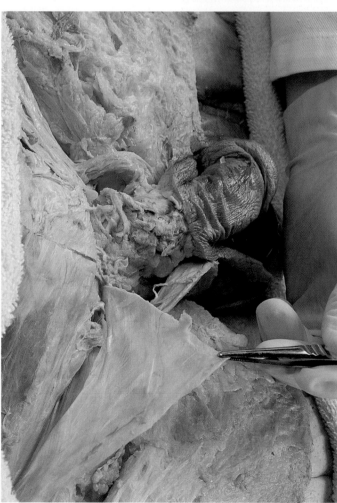

Finally, use a sharp knife to make a transverse section
through the testis and tunica vaginalis.

Observe:
○ *the **tunica vaginalis** (visceral and parietal layers),*
○ *the **tunica albuginea**,*
○ *the **septa** (septula),*
○ *the **seminiferous tubules**,*
○ *the **mediastinum testis**,*
○ *the **epididymis and its sinus**,*
○ *the **ductus deferens**.*

47

The Abdominal Viscera *in situ*

The peritoneal cavity will be opened and the viscera studied without further dissection. It is worth spending some time on this section, because the separate parts of the gastrointestinal tract will not be studied in order, from its proximal end to its distal end, in the following sections.

Make three incisions through the remaining layers of the abdominal wall (**1**):
- just above the inguinal ligaments, from one anterior superior iliac spine to the other,
- along the inferior margin of the thoracic cage, from one midaxillary line to the other,
- a vertical incision just to the left of the midline.

Turn the left flap laterally (**2**).

*Note that the liver is attached to the anterior abdominal wall by the **falciform ligament**.*

Divide the falciform ligament between the liver and the anterior abdominal wall, and turn the right flap laterally (**3**).

212 A
213 B

Identify the structures in the abdominal cavity, without disturbing them in any way.

Try to follow the parts of the gastrointestinal tract in the abdominal cavity sequentially from the stomach to the sigmoid colon. Find:
- *the **stomach**, feel the pylorus with your fingers,*
- *the first inch of the **duodenum** (the succeeding parts of the duodenum are not accessible at this stage).*
- *the **liver**, **gall bladder** and falciform ligament.*

Turn the greater omentum upwards and find:
- *the **jejunum**,*
- *the **ileum**,*
- *the **caecum** and **appendix**,*
- *the **ascending**, **transverse**, **descending** and **sigmoid** **colon**.*

Make a note of which segments of the GI tracts are suspended on mesenteries and which are directly attached to the posterior abdominal wall.

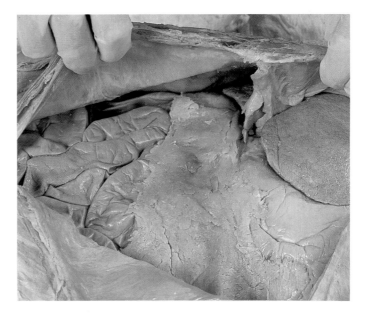

2

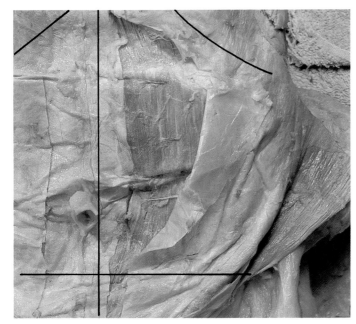

1

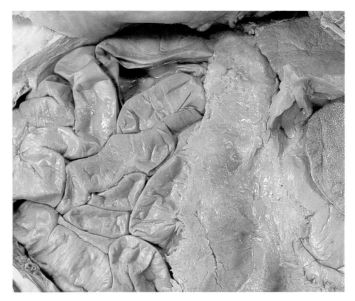

3

48

Exploration of the Peritoneal Cavity

The arrangement of the peritoneum and its reflections in the abdomen is difficult to understand. A good starting point is to realise which of the abdominal viscera are suspended from the posterior wall of the abdomen by mesenteries or their equivalent, but these arrangements will not be appreciated fully until the dissection of the abdomen is complete. By that time the peritoneum and its reflections will be destroyed, therefore repeat the exploration described in this section several times while you have the opportunity.

214 A

The abdominal cavity is divided by the **transverse mesocolon** into a **supramesocolic** and an **inframesocolic compartment**.
Lift up the **greater omentum** and find the **transverse colon** and the **transverse mesocolon** (**1**).
The supramesocolic compartment contains the liver, stomach and spleen; the inframesocolic compartment contains the small intestine and colon.

Return the greater omentum to its normal position and consider the supramesocolic compartment. This is subdivided into four **subphrenic spaces**.

Place your left hand on the right lobe of the liver and explore the **right anterior subphrenic space** (**2**).
It is limited by the **falciform ligament** medially, and the **superior leaflet of the coronary ligament** posteriorly.

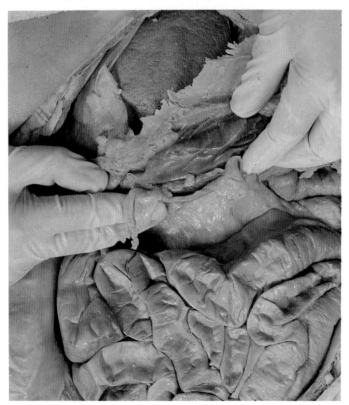

1

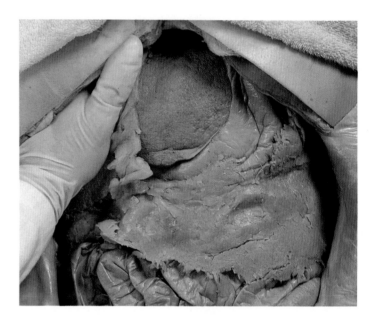

Explore the **left anterior subphrenic space** by placing your right hand on the left lobe of the liver; this space is limited by the falciform ligament and the superior leaflet of the coronary ligament (**3**).

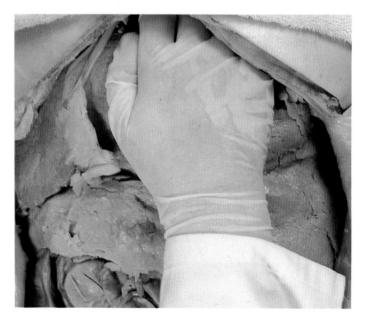

2

3

Put your hand around the left side of the stomach and feel for the spleen (4).

*The **left posterior subphrenic space** is usually called the **lesser sac of the peritoneum**.*

215 B

*To find the lesser sac of the peritoneum, first find the **epiploic foramen** leading into it. Feel for the **pyloric sphincter** at the distal end of the stomach. The epiploic foramen lies just superior, posterior and to the right of the pyloric sphincter. Place one or two fingers in the epiploic foramen and note its relations:*

217 C
- ○ *inferiorly: the first part of the duodenum,*
- ○ *anteriorly: a bundle of structures in the free edge of the lesser omentum,*
- ○ *superiorly: part of the liver,*
- ○ *posteriorly: the inferior vena cava, which feels hard because it contains a blood clot.*

Keep your fingers in the epiploic foramen and grasp the spleen with your right hand. The lesser sac lies behind the stomach and between your hands (5). Take note of its extent, walls and floor:
- ○ *anterior wall: lesser omentum and stomach,*
- ○ *floor: the transverse colon and transverse mesocolon,*
- ○ *posterior wall: the posterior abdominal wall and diaphragm.*

*The **right posterior subphrenic space** is also known as the **hepatorenal pouch** (of Morison). It is located between the liver, the right kidney and the hepatic flexure of the colon. It is the lowest part of the peritoneal cavity when the subject is in the prone position. Explore the right posterior subphrenic space (6).*

5

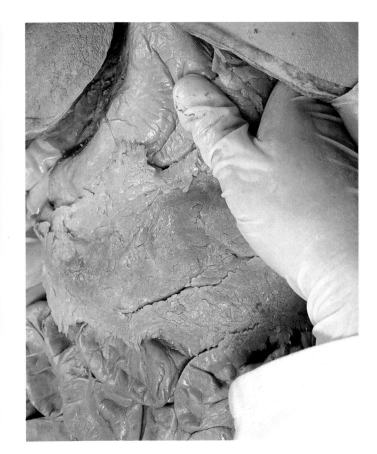

6

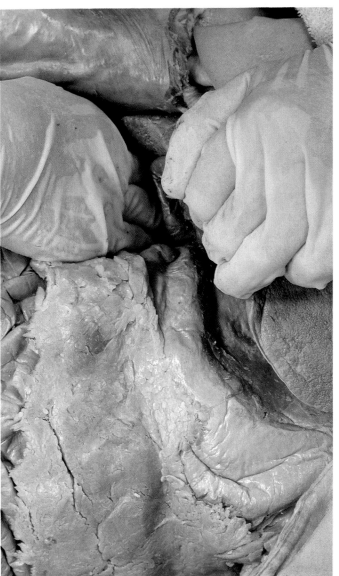

4

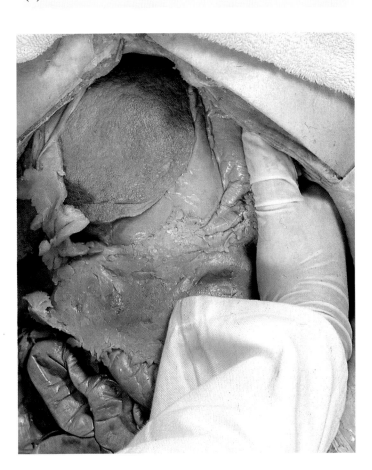

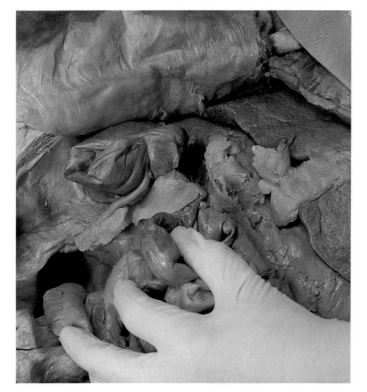

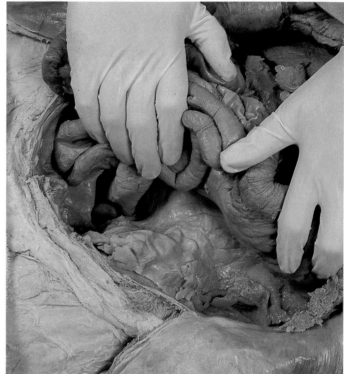

218 A

Lift up the greater omentum and transverse colon to explore the peritoneal arrangements in the inframesocolic compartment.

*Lift up the small intestine to find **the mesentery** running diagonally across the posterior abdominal wall from the upper left side to the lower right quadrant of the abdomen.*

*On the right side note two gutters running beside the ascending colon: the **medial** and **lateral paracolic gutters**.*
*The **lateral paracolic gutter** runs from the right posterior subphrenic space to the pelvic cavity. The right medial paracolic gutter does not communicate with the pelvic cavity; its inferior end is blocked by the mesentery*
Take this opportunity to identify:
○ *the **caecum**,*
○ *the **appendix**,*
○ *the **ileocaecal junction**,*
○ *the **peritoneal reflections** (folds) around these structures (**7**).*

Note the location of the appendix in your dissection. Move it around to simulate some of its other possible positions.

*There are **two paracolic gutters** on the left side, both of which communicate with the pelvic cavity (**8**).*
*Run your finger upwards in the left lateral paracolic gutter until you find a fold of peritoneum at the splenic flexure of the colon, **the phrenicocolic ligament**.*
*At the upper end of the left medial paracolic gutter observe the folds of peritoneum around the fourth part of the duodenum, and between these folds the **paraduodenal fossae**.*

49

The Small Intestine

It is easier to dissect the structures in the inframesocolic compartment first, to make the supramesocolic structures more accessible in later dissections. Here, the jejunum, ileum and mesentery are studied.

Turn the greater omentum upwards. Examine the jejunum and ileum once more *in situ*. Find the duodeno-jejunal and ileocaecal junctions.

Tie two ligatures around the distal end of the jejunum, close to its junction with the duodenum (**1**).

Cut between the two ligatures (**2**).

Tie a similar pair of ligatures around the distal end of the ileum, and cut between them (**3 & 4**).

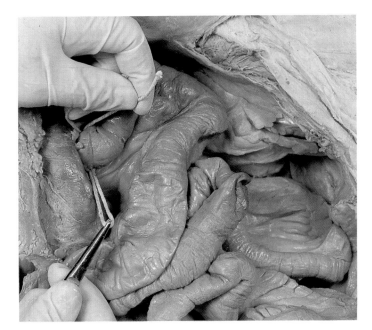

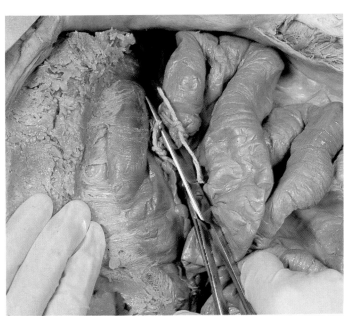

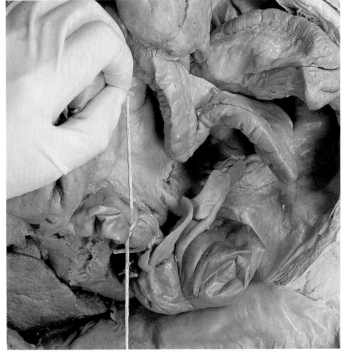

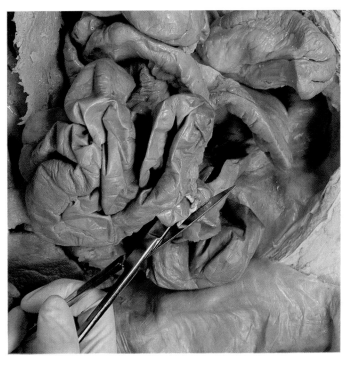

Cut through the mesentery approximately one inch in front of the abdominal wall (5).

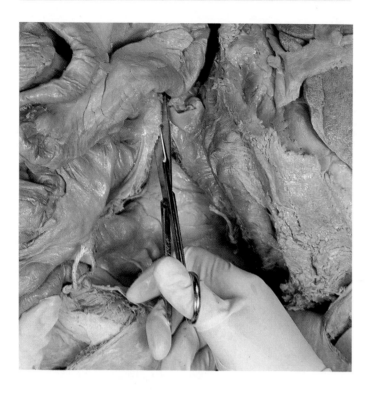

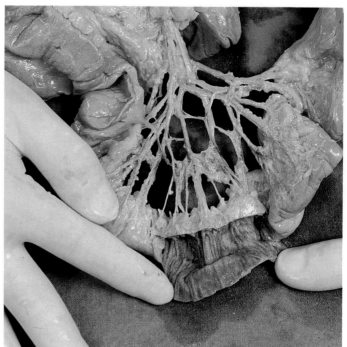

Compare the pattern of the blood vessels supplying the jejunum with that of the vessels supplying the ileum.

Compare the appearance of the mucosa of the jejunum with that of the ileum. Try to locate Peyer's patches (lymphoid tissue in the wall of the ileum).

Pick away the fat in the mesentery close to the jejunum, to observe the blood vessels which supply it. Cut a window in the wall of the jejunum and wash it out in running water (6).

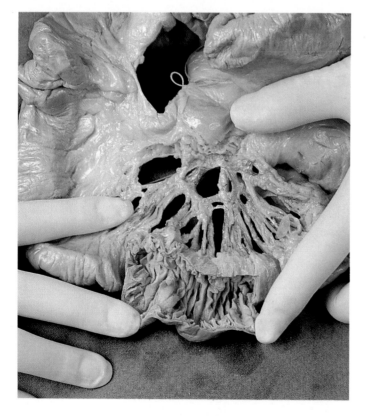

Clean the blood vessels in the mesentery and cut a window in the wall of the ileum (7).

50

The Colon

In this section the abdominal parts of the colon, blood supply and posterior relations of the colon will be studied.

Study the colon in situ (**1**).

*Find the **caecum** and **appendix**. Observe the **mesoappendix**, **vascular fold** and **bloodless fold** of the peritoneum.*

Incise the parietal peritoneum on both sides of the ascending colon, taking great care to avoid damaging the retroperitoneal structures.

Identify the following posterior relations of the ascending colon:
- *iliacus,*
- *quadratus lumborum,*
- *transversus abdominis muscles,*
- *the lower pole of the right kidney,*
- *the iliohypogastric nerve,*
- *the ilioinguinal nerve,*
- *the lateral femoral cutaneous nerve.*

*Find the **ileocolic** and **right colic arteries** under the peritoneum of the posterior abdominal wall, and trace these towards the mesentery.*

Separate the greater omentum from the transverse colon (use your fingers for this).

Identify the *transverse mesocolon* and divide it (**2**), leaving approximately one inch of it on the posterior abdominal wall. The structures lying posterior to the attachment of the transverse mesocolon will be studied with the duodenum and pancreas.

Tie a pair of ligatures at the junction between the descending colon and the sigmoid colon (**3**). Cut through the colon between the ligatures.

Carefully incise the peritoneum on both sides of the descending colon, and remove the entire colon.

1

2

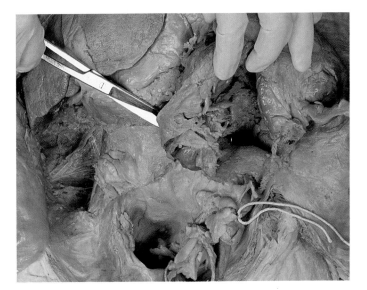

3

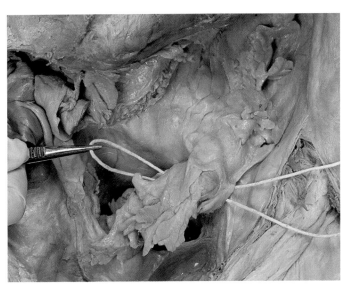

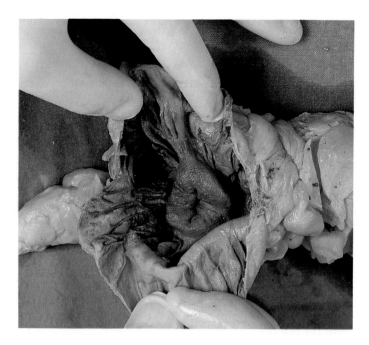

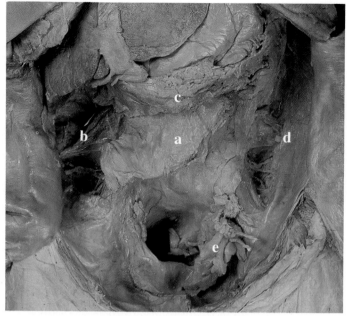

Identify the posterior relations of the descending colon:

- ○ *iliacus,*
- ○ *quadratus lumborum,*
- ○ *transversus abdominis muscles,*
- ○ *the lower pole of the left kidney,*
- ○ *the iliohypogastric nerve,*
- ○ *the ilioinguinal nerve,*
- ○ *the lateral femoral cutaneous nerve.*

Find the inferior mesenteric artery and its branches behind the peritoneum of the posterior abdominal wall.

a. **The mesentery**
b. **(Structures posterior to the ascending colon)**
c. **Root of the transverse mesocolon**
d. **(Structures posterior to the descending colon)**
e. **Sigmoid mesocolon**

Inspect the interior of the caecum.

Note the peritoneal reflections in the inframesocolic compartment (**5**).

Tie a ligature around the ascending colon, a few inches above the caecum. Open the caecum by means of a longitudinal incision along its right margin (**4**). Wash it out in running water.

51

The Stomach

In this section the stomach, its posterior relations and blood supply are studied with the stomach in place. Then the structures in free edge of the lesser omentum are defined. Finally, the stomach is removed and its interior is examined.

224 A

The inframesocolic viscera have been removed, which makes it easier to study the stomach, greater and lesser omenta and spleen in situ.

Use the atlas to help you to find:
- *the parts of the stomach,*
- *the lesser omentum,*
- *the greater omentum,*
- *the spleen.*

The diagram (1) may help you to identify the parts of the greater omentum:
- *the greater omentum proper, or gastrocolic ligament (a),*
- *the gastrosplenic ligament (b),*
- *the gastrophrenic ligament (c).*

Find also the lienorenal ligament between the spleen (lien) and the kidney.

Define the lesser sac of the peritoneum again, and note that the structures forming its posterior wall also form the 'bed of the stomach'.

The lesser sac may be entered through
- *the transverse mesocolon,*
- *the greater omentum, or*
- *the lesser omentum.*

As the transverse mesocolon was divided during the removal of the colon, push the stomach upwards and palpate the posterior wall of the lesser sac through the peritoneum to find (2):
- *the upper part of the abdominal aorta,*
- *the body and tail of the pancreas,*
- *the left kidney.*

2

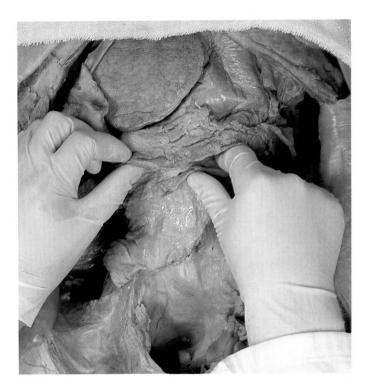

Carefully remove the peritoneum above the transverse mesocolon and clean the structures referred to above, as well as:
- the crura of the diaphragm,
- the splenic artery,
- the left gastric artery,
- the root of the hepatic artery.

220 A

Trace the three arteries back to the coeliac trunk, and try to identify the coeliac ganglia in the connective tissue surrounding the coeliac trunk.

Return the stomach to its usual position and revise the attachments of the lesser omentum to the stomach and to the liver.

1

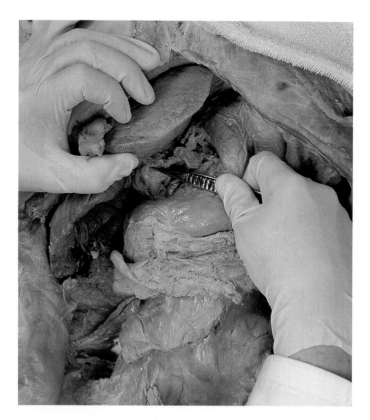

3

4

Clean the structures in the free edge of the lesser omentum (**3**).

> *Identify:*
> ○ *the **portal vein**,*
> ○ *the **hepatic artery**,*
> ○ *the **bile duct**.*

220 A

Clean the arteries which run along the greater (**4**) and lesser curvatures of the stomach. Trace them back to their origins.

> *Use the atlas to identify the arteries which supply the stomach.*

To remove the stomach:
- separate the *nerves* surrounding the distal end of the oesophagus by blunt dissection, with closed forceps,
- divide the *abdominal oesophagus* as close as possible to the diaphragm, leaving at least some of the nerves *in situ*,
- divide the *gastrosplenic* and *gastrophrenic ligaments* close to the stomach, but leave the arteries with the stomach,
- find the *pylorus* by palpation and divide the *duodenum* just distal to the *pyloric sphincter*,
- lift out the stomach and its blood vessels.

Open the stomach over a sink, by means of a longitudinal incision in its anterior surface, parallel to the greater curvature. Wash out the stomach in running water.

> *Examine the contents and mucous membrane of the stomach.*

52

The Liver

*In this section the spaces around the liver are studied, and the dissection
of the extrahepatic biliary apparatus is completed. Finally, the liver is
removed from the body for further examination.*

*Review the spaces around the liver. The **hepatorenal
pouch** (of Morison) is much easier to demonstrate
now that the infrahepatic viscera have been removed.*

Find the *gall bladder*. Remove the peritoneum which
covers it and clean the gall bladder, cystic duct and
cystic artery. Follow the duct system towards the *porta
hepatis*.

*Draw a diagram of the extrahepatic biliary apparatus
in your dissection, and compare your findings with
the descriptions given in the standard textbooks of
anatomy.*

Take great care to remove the liver so that its posterior
relationships can be appreciated:
- cut through the *falciform ligament* using scissors,
- identify the superior leaflet of the *coronary ligament*
  and the *two triangular ligaments* with the tips of your
  fingers,

- cut through the superior leaflet of the coronary
  ligament with the tip of a scalpel blade,
- let the liver fall downwards,
- identify the inferior leaflet of the coronary ligament,
- cut through the inferior leaflet with a scalpel.

The liver is now attached to the body by blood vessels:
- divide the *hepatic artery* close to the structures that
  were running the free edge of the lesser omentum,
- cut through the *inferior vena cava*, just below the
  liver,
- free the inferior vena cava from the central tendon
  of the diaphragm from the thoracic aspect.

Lift out the liver.

*Use the atlas to help you to identify the features of the
inferior and diaphragmatic surfaces of the liver.
Identify the impressions left by the abdominal organs
related to the liver* (**1**).

227 A
228 B

1

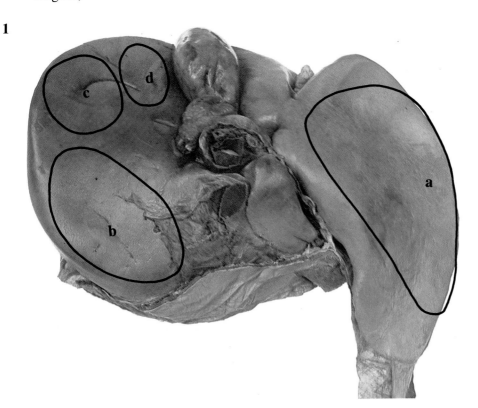

a. **Gastric impression**
b. **Renal impression**
c. **Colic impression**
d. **Duodenal impression**

53

The Duodenum, Pancreas, Spleen and Portal Vein

In this section all these organs will be studied and then removed as a block so that their posterior relations may be studied.

Before starting to dissect identify:
○ *the* **spleen**,
○ *the* **pancreas**,
○ *the four parts of the* **duodenum** *by palpation.*

Locate both **kidneys** *by palpation.*

Take note of the relations of the **root of the transverse colon** *to all these structures.*

Remove the peritoneum from the anterior surface of the structures listed above. Try to avoid damage to the arteries and ducts associated with them. Leave the root of the transverse mesocolon in place if you can.

225 B
Identify:
○ *the* **head, body, tail** *and* **uncinate process** *of the pancreas,*
○ *the main trunk of the* **superior mesenteric artery**.
Find the **abdominal aorta** *by palpation. Note the relations of the superior mesenteric artery to the parts of the pancreas, to the duodenum and aorta.*

Clean the superior mesenteric artery in the inframesocolic compartment, and demonstrate as many of its branches as possible. Preserve the companion veins and at least some of the lymph nodes in the mesentery (**1**).

Identify:
○ *the* **inferior pancreaticoduodenal artery**,
○ *the* **middle colic artery**,
○ *some jejunal and ileal brances,*
○ *the* **right colic artery**,
○ *the* **iliocolic artery**,
○ *the* **superior mesenteric vein**.

Return to the supramesocolic compartment and separate the structures in the free edge of the lesser omentum.
Trace the artery, vein and duct upwards towards the liver and downwards towards the duodenum and pancreas.

Identify:
○ *the* **portal vein**,
○ *the* **hepatic artery**,
○ *the* **bile duct**.

Clean the *splenic vein* and the *superior mesenteric vein*, and follow them to the right and upwards until they join together to form the *portal vein*.

Lift up the four parts of the duodenum and head of the pancreas, and clean behind them to determine their posterior relations (**2**).

1
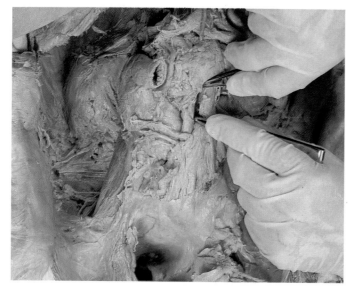

2
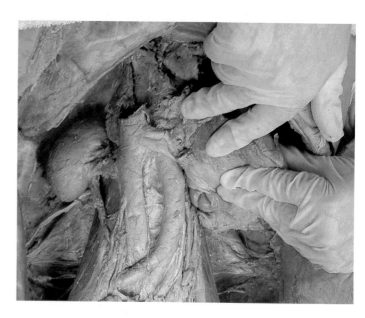

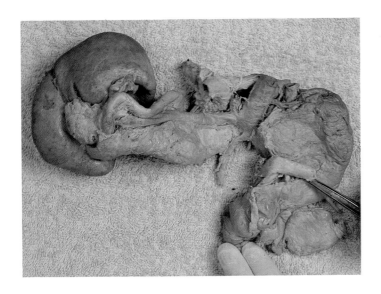 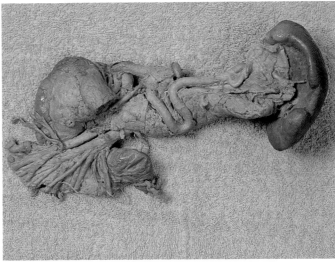

Study the posterior relations of the duodenum and head of the pancreas.

Lift up the spleen and tail of the pancreas, and clean the area behind them to display their posterior relations.

Study the posterior relations of the tail of the pancreas.

Divide:
- the structures in the free edge of the lesser omentum close to the duodenum.
- the splenic artery and vein close to their proximal ends,
- the superior mesenteric vessels close to the head of the pancreas.

Remove a block of tissue consisting of the duodenum, pancreas, spleen and their associated vessels.

Inspect the anterior surface of this specimen (3).

Clean the *bile duct* and follow it downwards. Demonstrate the *pancreatic duct* by blunt dissection. Follow the bile duct and pancreatic duct to the duodenum.

Look at the portal vein and its tributaries (4).

Open the duodenum by means of a longitudinal incision in its posterior wall. Wash it out under running water.

*Observe the way in which the bile and pancreatic ducts enter the duodenum. Inspect the mucous membrane of the duodenum, try to find the **duodenal papilla**. Compare your findings with the descriptions given in the standard textbooks.*

225 C

Replace the block containing the duodenum, pancreas and spleen and check the posterior relations once more.

The Kidneys and Ureters

The purpose of this section is to demonstrate the shape, size and position of the kidneys and ureters on the posterior abdominal wall, with an emphasis on their posterior relations. This will help you to understand the radiological appearance of these organs. In addition, the adrenal glands and gross internal anatomy of the kidney will be studied.

Locate the kidneys on the upper part of the posterior abdominal wall by palpation through the peritoneum.

Remove the peritoneum, perirenal fat, and the fat and connective tissue around the ureter. It is best to use your fingers for this task, removing the fat in large pieces. Be careful to preserve the *adrenal glands* situated near the upper poles of the kidneys.

Clean the vessels entering the hila of both kidneys.

236 A

Observe the precise relationships between the kidney and adrenal gland on both sides.

Observe the arrangement of the vessels at the hila on both sides.

Observe the position of the ureter on the posterior abdominal wall on both sides.

Pull the kidney gently away from the posterior abdominal wall and turn it medially (**1**).

1

Clean the area which was just posterior to the kidney, and define the nerves and muscles in this region. You should be able to do this with your fingers.

Observe the posterior relations of the kidney:
○ *the **twelfth rib**,*
○ *the **psoas muscle**,*
○ *the **quadratus lumborum muscle**,*
○ *the **subcostal nerve**,*
○ *the **iliohypogastric nerve**,*
○ *the **ilioinguinal nerve**.*

Replace the kidney to its normal position and check its posterior relations for the second time.

With a sharp knife, cut through the kidney in the coronal plane, starting at the convex border and ending at the hilus.

Inspect the cut surface and identify as much as you can, using the illustration in the atlas.   **238 A**

The Posterior Abdominal Wall

In this section the entire posterior abdominal wall will be cleaned and studied systematically using the illustrations in the atlas. Then the viscera which have been removed in earlier stages of the dissection will be carefully replaced so that you can review their relationships to each other.

Clean the entire posterior abdominal wall. Use your fingers for this or blunt forceps. Use a scalpel only if absolutely necessary (**1**).

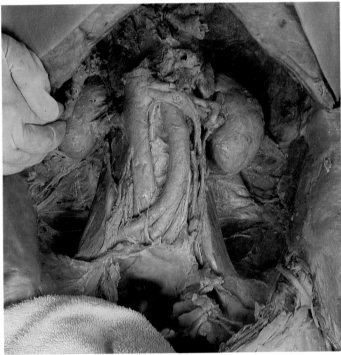

1

241 C *Use the atlas to study the structures which form the posterior wall of the abdomen.*
Diaphragm, *find:*
○ *the **crura**,*
○ *the **arcuate ligaments**,*
Psoas major: *study its anterior lateral and medial*
241 C, *relations with care. Identify all the nerves (branches*
D *of the lumbar plexus) that appear on its medial, lateral and anterior surfaces.*
Note the relations of the gonadal vessels to the ureters, to each other and to the psoas major muscle at various levels.

Quadratus lumborum, *note the nerves on its anterior*
241 D *surface. Define its lateral edge, push your finger tips under this edge to feel the **thoracolumbar fascia** over the erector spinae muscle, and confirm that this layer of the thoracolumbar fascia gives origin to the internal oblique muscle.*

The great vessels of the abdomen: **236 A**
○ *study their branches and relations,* **242 A**
○ *Try to find the **cisterna chyli** behind the upper part of the abdominal aorta.*

*Replace the block containing the duodenum, pancreas and spleen on to the posterior abdominal wall in its proper position (**2**). This will help you to remember the relations of these organs.*
Replace:
○ *the liver*
○ *the stomach*
○ *the colon*
in their proper positions on the posterior abdominal wall, taking into account all you have learned about their relations in the preceding sections.

2

Osteology and surface anatomy of the Abdomen

A thorough knowledge of the surface anatomy of the abdominal organs and vessels is obviously of great importance for the practising clinician in most branches of medicine. It will be very helpful if this section is supplemented with the study of as many radiographs of the abdomen as you can find.

Bony landmarks are identified:
- on the articulated skeleton,
- in suitable radiographs,
- by palpation on the living subject.

The posterior superior iliac spine is difficult to palpate, but it can often appear as a small dimple (**j**).

It may be difficult to palpate the twelfth rib which sometimes does not project beyond the erector spinae muscles (**2**).

Anterior abdominal wall: skeletal landmarks

a. Xiphoid process
b. Costal margin
c. Subcostal angle
d. Iliac crest
e. Tubercle of the iliac crest
f. Anterior superior iliac spine
g. Pubic crest
h. Pubic tubercle

1

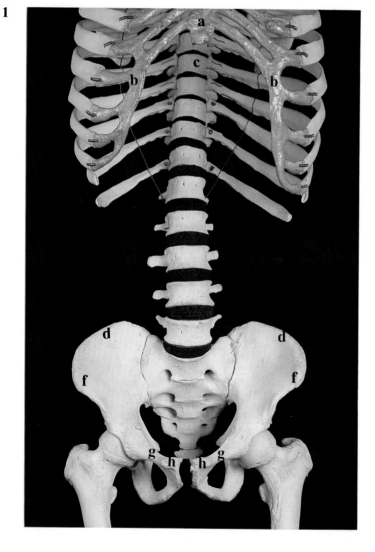

2

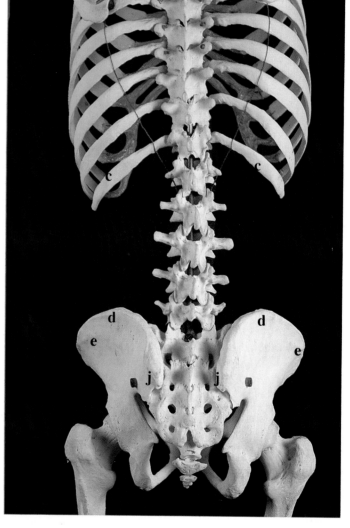

On the living subject indicate (**3**):
- the inguinal ligament, between the anterior superior iliac spine and the pubic tubercle,
- the costal margin, which you have just located by palpation,
- the domes of the diaphragm in quiet inspiration: the right dome in the fifth intercostal space, and the left dome by the sixth rib in the midclavicular line.

3

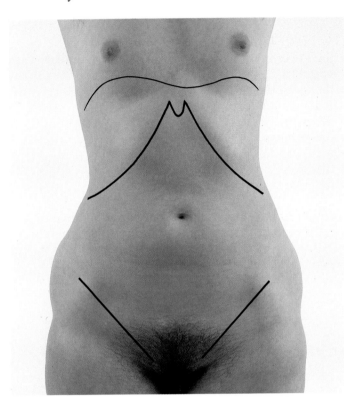

Look for the linea alba and lineae semilunares in a thin or muscular subject (**4**).

Now find the surface characteristics of several important markings.

4

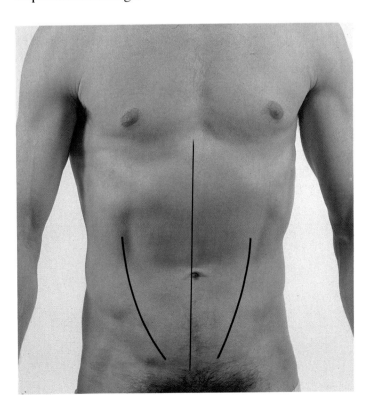

Liver
On the left side, draw a line from the highest point of the dome of the diaphragm diagonally downwards, to the lowest part of the costal margin on the right. Most of the liver lies within the thoracic cage.

Gall bladder
This is located on the right side of the body, where the linea sculunans meets the costal margin. This is also the posterior of the tip of the ninth costal cartilage.

Stomach
The stomach lies to the left of the liver, but its position varies depending on habitus, whether the body is erect or supine, and whether the stomach is full or empty. Therefore, there is not much point in outlining it. The position of the oesophageal opening is more or less constant, one inch to the left of the xiphoid process (**5**).

5

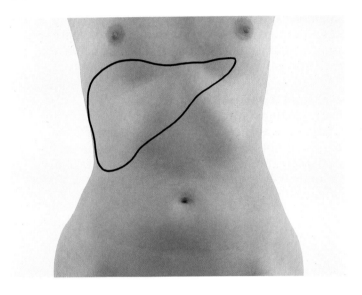

Appendix
McBurney's point is supposed to correspond to the normal position of the appendix. It is located one-third of the way along a line joining the anterior superior iliac spine with the umbilicus (**6**).

6

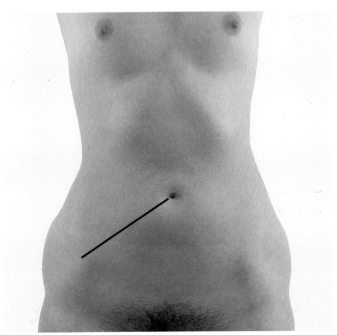

Spleen

The spleen lies between the ninth, tenth and eleventh ribs, with its long axis along the tenth rib. Its lateral end is in the mid-axillary line, and its medial end is approximately two inches from the (posterior) midline.

Kidneys

Draw rectangles on the posterior abdominal wall; the horizontal lines:
- at the level of T1 approximately,
- at the level of L4 approximately,

the vertical lines:
- four centimetres and ten centimetres from the midline.

Outline the kidneys within these rectangles, with the right kidney one inch or so below the left (7). Radiographs will confirm that positions of the above organs are quite variable.

Small intestine and colon

Refer to radiographs to appreciate the landmarks and usual locations of the small intestine and colon, and keep in mind their locations in the living subject.

Regions of the abdomen

The abdomen is often divided into nine regions by reference to vertical and horizontal planes. This division is not very helpful in understanding the topography of the abdominal viscera, but variants of abdominal divisions are often used for descriptive purposes.

The *horizontal planes* are divided into:
- The transpyloric plane, quite useful in visceral topography. It usually lies between the top of the pubic symphysis and the jugular notch, and approximately halfway between the xiphoid process and the umbilicus (8).

- The transtubercular plane, between the tubercles of the iliac crest.

The *vertical planes* are the midclavicular lines.

The transpyloric plane corresponds to:
- the lower margin of the body of the first lumbar vertebra,
- the tip of the cartilage,
- the gall bladder,
- the pylorus,
- the kidneys,
- the pancreas,
- the artery,
- the lowest extent of the spinal cord (8).

Another important landmark is the mid-inguinal point, located halfway between the pubic symphysis and the anterior superior iliac spine. This corresponds to the femoral artery.

7

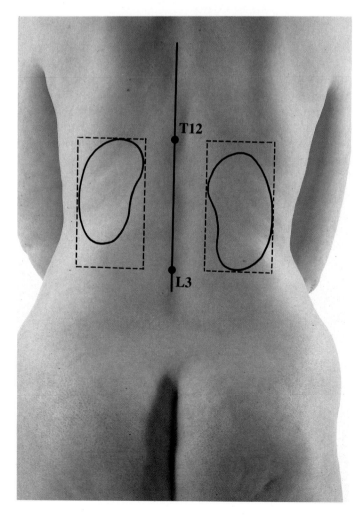

8

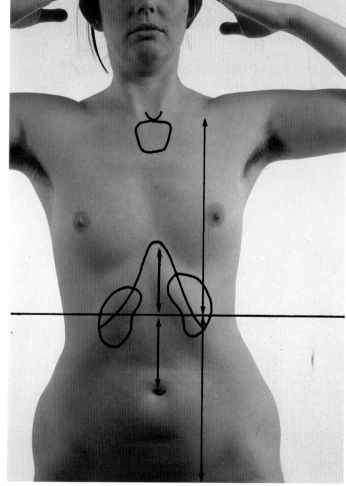

Pelvis and Perineum

The Pelvic Viscera *in situ* in the Male and Female

The purpose of this dissection is to locate the inlet to the pelvis, note the disposition of the rectum, uterus and bladder, and observe the relation of the peritoneum to these structures. Ensure that the viscera in situ *are studied in both sexes.*

Male

250 A

*Look at the pelvic inlet. Find the **promontory of the sacrum** and **pubic symphysis**. Follow the pelvic brim around between these landmarks. Note that the pelvic inlet faces forwards.*

*Locate the position of the **urinary bladder**; this lies beneath the peritoneum, anteriorly.*

*Find the **rectum** posteriorly; the **sigmoid colon** leads into it.*

*Explore the **rectovesical pouch** of peritoneum, between the rectum and the bladder.*

Note the peritoneal relations of the rectum in the pelvis:
○ *the front and sides of the upper part are invested by peritoneum,*
○ *only the front of the inferior part is covered by peritoneum.*

Strip off the peritoneum around the pelvic brim.

242 A

Identify the structures which cross the pelvic brim or are close to it:
○ *the **common and external iliac arteries**,*
○ *the **internal iliac arteries**,*
○ *the **lumbosacral trunk**,*
○ *the **obturator nerves**,*
○ *the **ureters**,*
○ *the **sigmoid colon and its mesocolon**,*
○ *the **superior rectal branches** of the **inferior mesenteric vessels**,*
○ *the **vasa deferentia**.*

Female

Examine the pelvic inlet and note that it faces forwards.

*Find the **rectum, uterus** and **urinary bladder** (1).*

*Explore the **rectouterine pouch** (of Douglas) between the rectum and the uterus. Explore the **uterovesical pouch** between the uterus and the bladder.*

*Observe the **fundus of the uterus**, and the **broad ligament** between the uterus and the pelvic wall. Find the **uterine tubes** in the free edge of the broad ligament, and the **infundibulum** and **fimbria** around their openings into the peritoneal cavity.*

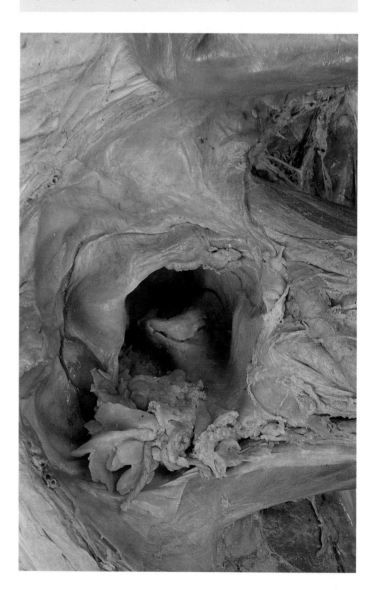

1

*Find the **ovaries** on the posterior surface of the broad ligament, and note their relations on the wall of the pelvis (2).*

*Identify the structures listed in the dissection of the male pelvis, as they cross the pelvic inlet. The **round ligaments of the uterus** occupy the positions taken by the vasa deferentia in the male, but it may be difficult to demonstrate them.*

Remove the peritoneum near the pelvic brim.

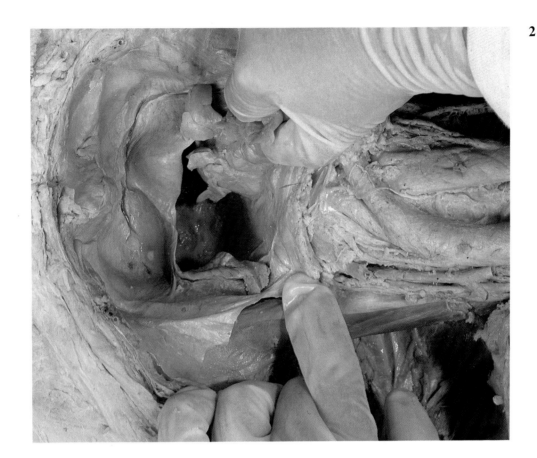

2

58

The Sagittal Sections of the Pelvis in the Male and Female

*The pelvis will be sectioned in the sagittal plane, and the pelvic viscera
will be studied without further dissection.*

In both sexes

- Cut through the intervertebral disc between the third and fourth lumbar vertebrae, and use a knife to cut through the muscles of the body wall (**1**).

- Use a scalpel to divide the soft tissues and pubic symphysis in the midline (**2**).

- Turn the body over, and use a saw to cut through the sacrum in the midline (**3**).

Wash out both halves of the specimen in running water.

1

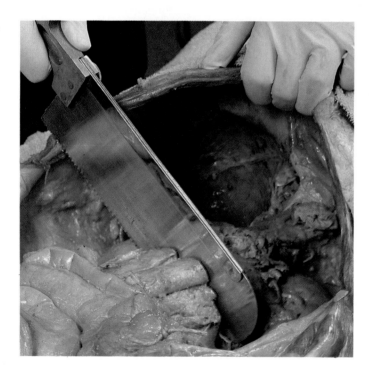

2

3

Male

Look at the interior of the bladder, and note the positions and relations of the prostate gland. **248 A**

*Identify the parts of the rectum, and note their positions with respect to the sacrum and coccyx. Note that the retrovesical pouch is not deep, and inferiorly the rectum and bladder are separated by the **rectovesical fascia**.*

*Disturb the rectovesical fascia slightly, to find the **seminal vesical vas, ducturs deferens** and **ejaculatory ducts** just posterior to the bladder and above the prostate gland.*

Follow the peritoneum from the anterior abdominal wall, across the pelvic viscera, to the front of the rectum.

Place a finger in the rectum (4), and note the structures which can be palpated through the walls by rectal examination.

4

Female

254 A

*Look at the rectum, uterus and bladder.
Follow the peritoneum from the anterior abdominal wall across these organs.*

*Note the **fundus** and **cervix of the uterus**.*

Once again, look at the structures on the broad ligament (5).

5

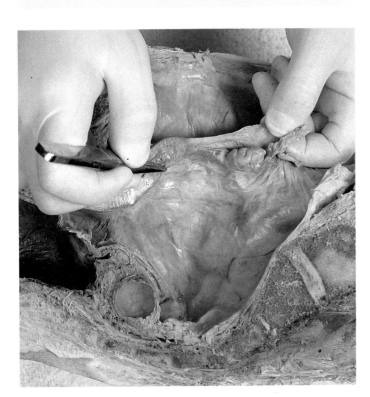

*Observe the **vagina** in the perineum and pelvis.
Demonstrate the **anterior**, **lateral** and **posterior fornices of the vagina**.*

Demonstrate the very close relationship between the posterior fornix of the vagina and the peritoneum in the floor of the rectouterine pouch.

Place your finger in the vagina (6), and observe which structures can be palpated by vaginal examination.

Similarly, observe the structures palpable by rectal examination in the female.

6

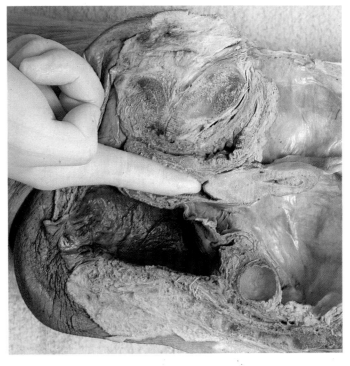

The Pelvic Walls and Vessels

The pelvic walls are lined by muscles, two of which (obturator internus and piriformis) will not be fully appreciated until the lower limb has been studied. The levator ani muscle forms the floor of the pelvis or pelvic diaphragm. Only half of it can be seen here.

The dissection of the vessels will be confined to demonstrating the branches of the internal iliac artery; the veins form complex systems of plexuses in the pelvis.

- Use one half of the pelvis only, retaining the best half for revision.

- Pull the rectum medially (**1**) and remove it. Divide it just above the anal canal.

- Remove most of the bladder, retaining only the posterior wall and *trigone*.

- Retain the *prostate gland* in the male, and the *uterus* and *broad ligament* in the female.

- Carefully remove the rest of the peritoneum in the pelvis.

- Clean all the branches of the internal iliac artery (**2**).

2

1

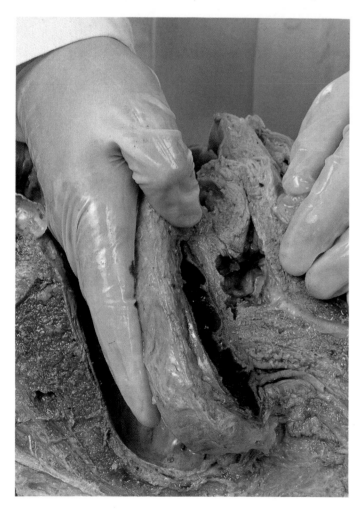

*Identify the **branches of the internal iliac artery** by tracing them forwards to the organ(s) they supply, or out of the pelvis.*   **251 C**

*Find the **ischial spine** by palpation.*

*Follow the **piriformis muscle** laterally from its origin on the sacrum, through the greater sciatic notch, above the ischial spine. Do not clean it at this stage.*   **245 A**

*Pull medially the remains of the bladder and prostate gland or the uterus, to display the superior surface of **levator ani**. This muscle is quite thin, therefore do not try to clean it.*

Define the superior attachment of levator ani; try to visualize it as a bowl-shaped muscle supporting the bladder, prostate and rectum in the male, or the bladder, uterus and rectum in the female.   **245 A,B**

60

The Lumbosacral Plexus

The main branches of the lumbar plexus have been seen in the dissection of the posterior abdominal wall. The lumbosacral trunk was displayed as it crossed the pelvic brim.

In this dissection the sacral ventral rami are cleaned on the surface of the piriformis muscle, and traced towards the greater sciatic notch through which they leave the pelvis to enter the lower limb.

Identify the piriformis and greater sciatic notch by palpation.

Clean the five roots of the sacral plexus (**1**).

Using the illustrations in the atlas, identify the roots and major branches of the sacral plexus.   **252 A**

The sacral plexus will be studied further in the dissection of the gluteal region of the lower limb.

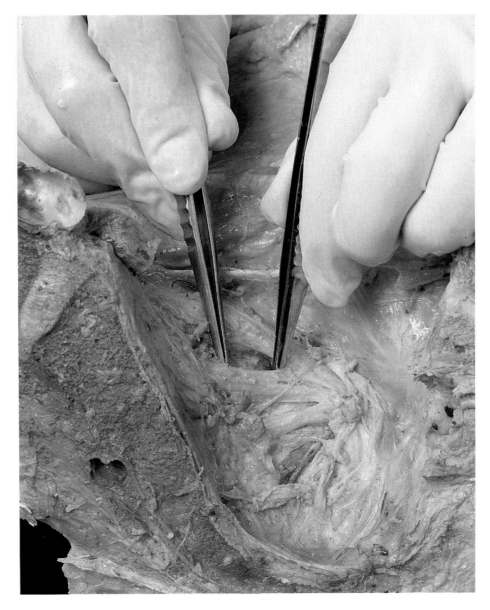

1

The Perineum in the Male and Female

First, the perineum is located in the sectioned pelvis. The extent of the perineum is then demonstrated on an articulated pelvis. Returning to the dissection, the ischiorectal fossa is shown in the bisected pelvis. This will also demonstrate the position of the urogenital diaphragm. Finally, the structures on the inferior surface of the urogenital diaphragm will be demonstrated in the male.

The general arrangement of the perineum in the female is similar to that in the male. However, its dissection is much more difficult as the urogenital diaphragm is pierced by the vagina and urethra; also, the structures of the inferior surface of the urogenital diaphragm are much smaller in the female. Concentrate on the male specimen, to help you understand the anatomy of this region.

Male

*On the undissected half of the pelvis demonstrate the position of the perineum below the pelvis (**1**). Note that it includes the penis and anal canal.*

Follow the male urethra from:
○ *the bladder,*
○ *through the **prostate gland (prostatic urethra)**;*
○ *through the **urogenital diaphragm (membranous urethra)**;*
○ *through the **bulb** and **corpus spongiosum** of the penis **(penile urethra)**.*
Note the double curve of the urethra between the bladder and its orifice at the tip of the penis.

*Examine the **anal canal**, and note the angle between the anal canal and rectum.*

It is important to have a clear idea of the three-dimensional arrangement of the perineal membrane and ischiorectal fossae.

*On the articulated pelvis find the **ischiopubic rami** (**2**). The **urogenital diaphragm** extends between the ischiopubic rami (**a**). In the male, this is pierced only by the (membranous) urethra. Note also the **anal triangle** (**b**).*

1

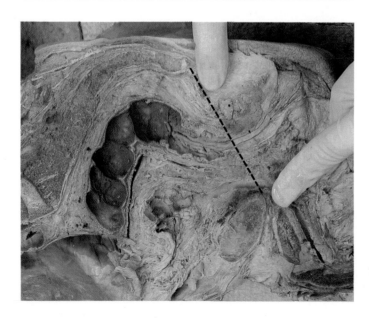

2

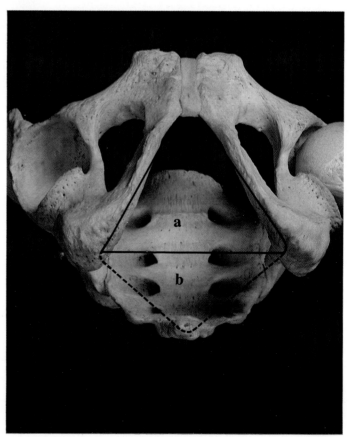

Review the attachments of levator ani.

The perineum is located inferior to the levator ani muscle, the pelvic diaphragm. The ischiorectal fossae fill the perineum lateral to the levator ani and anal canal, and extend anteriorly above the perineal membrane.

Make an incision just lateral to the anus in your specimen.

Remove some of the loose fat beside the anal canal.

*Place your finger into the incision and explore the space that previously contained the fat, the **ischiorectal fossa** (3). By pushing laterally you should be able to feel the **obturator internus muscle**. You may also feel the **neurovascular bundle** against the muscle. This contains the **pudental nerve** and **internal pudental vessels**.*

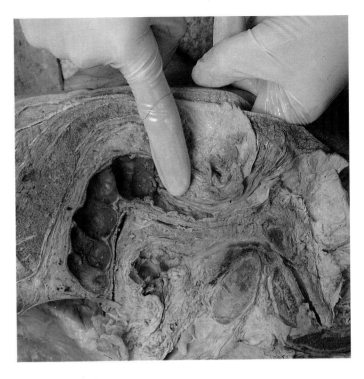

4

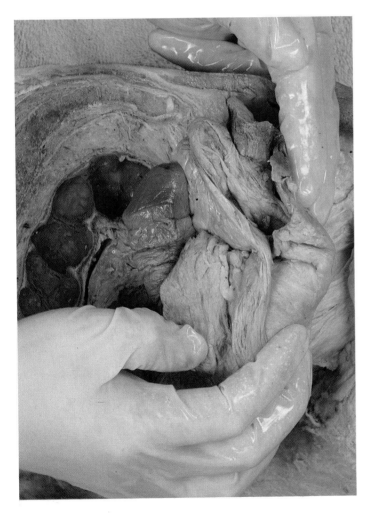

3

*Feel the roof of the **ischiorectal fossa**; this is the inferior surface of **levator ani**. Note that the ischiorectal fossa extends from the level of the anal canal to the level of the rectum (4).*

*Push your fingers forwards into the anterior recess. The floor of this recess is the **urogenital diaphragm**, or **perineal membrane**.*

Clean the structures located in the superficial perineal pouch, that is, on the inferior surface of the perineal membrane.

Use the illustrations in the atlas to identify these structures.

256 B,C

It is difficult to dissect this region in more detail, but once the fundamental arrangement of the urogenital diaphragm, ischiorectal fossae and superficial perineal structures is appreciated, it will be easier to understand descriptions of this region given in the standard textbooks of anatomy.

Female

Demonstrate the perineal structures on the undissected half of the pelvis.

Note the length of the urethra.

Note the location of the vagina in the pelvis and perineum.

Find the perineal body between the vagina and the anal canal.

Make an incision through the skin just lateral to the anal canal.

Remove some of the loose fat beside the anal canal.

Explore the ischiorectal fossa with your finger. It should be possible to demonstrate the obturator internus and levator ani muscles as well as the neurovascular bundle, but the anterior recess is more difficult to observe properly.

257 E

62

Osteology of the Pelvis

The main purpose of this section is to enable you to appreciate the anatomical position of the articulated pelvis, and to relate its various parts to the abdomen, pelvis, perineum and lower limb.

*Study the bony pelvis in an articulated skeleton. It consists of the two **hip bones**, **sacrum** and **coccyx**.*

In the anatomical position, the two anterior superior iliac spines and upper margin of the pubic symphysis are in the same coronal plane.

*Trace the superior aperture of the pelvis with your finger (**1**):*
a. sacral promontory
b. arcuate line
c. iliopubic eminence
d. pectineal line
e. pubic crest
f. upper margin of the pubic symphysis

The perineum lies below the pelvic floor. Run your fingers around the inferior aperture of the pelvis:
○ *the **inferior margin of the pubic symphysis**,*
○ *the **ischiopubic ramus**,*
○ *the **ischial tuberosity**,*
○ *the posterolateral margin between the ischial tuberosity and the coccyx (position of the sacrotuberous ligament).*
These structures also form the boundaries of the perineum.

Now relate the lesser pelvis to the regions of the lower limb:
○ *the **gluteal region**: the pelvis communicates with the gluteal region through the **greater sciatic notch** (**2**),*
○ *the **front of the thigh**: the pelvis communicates with the front of the thigh through the interval below the inguinal ligament (**1**),*
○ *the **medial compartment** of the thigh: the pelvis communicates with the medial compartment of the thigh through the **obturator canal** (**1**).*

*The **lesser sciatic foramen** leads into the perineum.*

1

*Below this aperture lies the 'true' or **lesser pelvis**. Above this aperture lie the iliac fossae and posterior wall of the abdomen.*

*Use the illustrations in the atlas to learn the details of the inner surfaces of these bones. Pay particular attention to the attachments of **levator ani** and **coccygeus muscles**, to understand how these muscles form the pelvic floor.*

*Reconstruct the positions of the **sacrotuberous** and **sacrospinous** ligaments; define the **greater** and **lesser sciatic foramina**.*

301 C

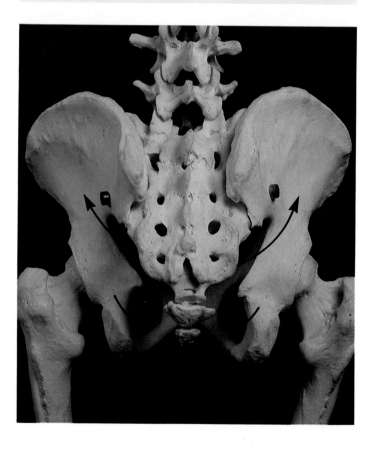

2

Lower Limb

63

The Femoral Triangle and Canal

*In this dissection the most important structures are the femoral artery,
femoral nerve and great saphenous vein. When these structures are
demonstrated, the muscles of the anterior compartment of the thigh will
be identified; details of the origins and insertions of these muscles are left
for a later dissection.*

Make a longitudinal incision on the medial aspect of
the thigh, and a transverse incision just below the
patella. Reflect the skin flap from the medial, anterior
and lateral aspects of the thigh (**1**).

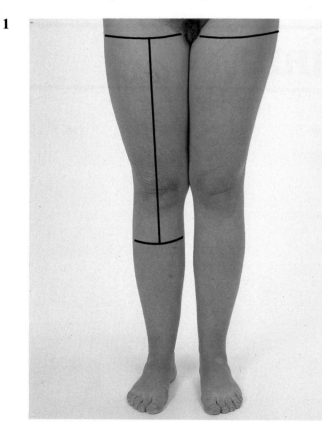

1

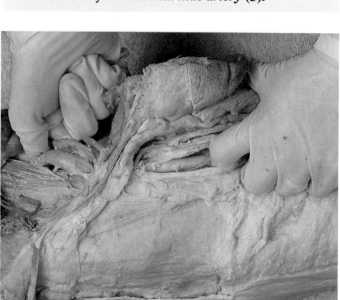

2

Open the *femoral sheath* using scissors, taking care to
avoid cutting any of its contents. Clean the *femoral
artery* and *femoral vein*.

Expose the *femoral nerve* as it enters the thigh below
the middle of the inguinal ligament.

> *Demonstrate the **branches of the femoral nerve**,
> **branches of the femoral artery** and tributaries of the
> femoral vein* .
>
> *Show that, in the thigh, the femoral artery is a
> continuation of the **external iliac artery** (**3**).*

**296 A,
B
297 D**

3

*Identify the **great saphenous vein**, running in the fatty
superficial fascia along the medial part of the thigh.
Trace its full extent in the thigh (**2**).
Identify the groups of **superficial inguinal lymph
nodes**.*

*Follow the great saphenous vein to the **cribriform
fascia**. Demonstrate the falciform margin of the
saphenous opening using a finger or blunt forceps to
remove the cribriform fascia.
Open the great saphenous vein to demonstrate its
valves.*

*Define the boundaries of the **femoral triangle**. Use
your fingers to define the **femoral sheath**. Take careful
note of the **relations of the femoral sheath** .*

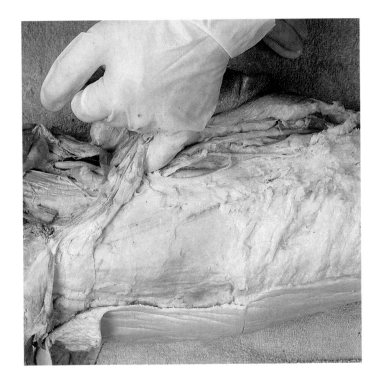

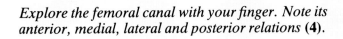

Explore the femoral canal with your finger. Note its anterior, medial, lateral and posterior relations (4).

Carefully remove the superficial fascia, preserving the great saphenous vein and lymph nodes, to expose the *fascia lata.*

Observe the cutaneous nerves of the thigh, which lie on the fascia lata.

*Identify the **iliotibial tract**. Follow the iliotibial tract from the **iliac crest** to the **lateral condyle of the tibia**.*

Make a longitudinal incision in the anterior and upper part of the fascia lata, and reflect flaps to display the *tensor fasciae latae muscle.*

Make a longitudinal incision through the fascia lata in the midline of the thigh, from the inguinal ligament to a point just above the patella. Reflect flaps of fascia lata laterally and medially.

Identify and clean the entire length of the *sartorius muscle*. Define its upper and lower attachments.

298 A
299 B

*Lift the **sartorius muscle**. Use your fingers or blunt forceps to demonstrate the walls and contents of the **adductor canal** (5).*

298 A

*Move the sartorius muscle in order to examine the origins and insertions of the **four parts** of the quadriceps femoris muscle. Use your fingers to separate the four parts of this muscle. Make a longitudinal incision over the patella, and use a probe to explore the **prepatellar bursa**.*

299 C,D

The Medial Aspect of the Thigh

In this dissection the muscles of the medial (adductor) compartment of the thigh are studied together with the nerve of the medial compartment, the obdurator nerve.

Remove the superficial and deep fascia from the medial aspect of the thigh.

Move the quadriceps femoris muscle laterally, to display the *muscles of the adductor compartment.*

297 C

Identify the muscles of the adductor compartment:
○ *adductor longus,*
○ *adductor brevis,*
○ *adductor magnus,*
○ *gracilis.*
Find the intermuscular septum between the anterior and medial compartment of the thigh .

Use your fingers to separate these muscles from each other (**1**).

Find the branches of the obturator nerve running between the muscles.

*Trace the **femoral vessels** through the opening of the **adductor magnus muscle**.*

297 C

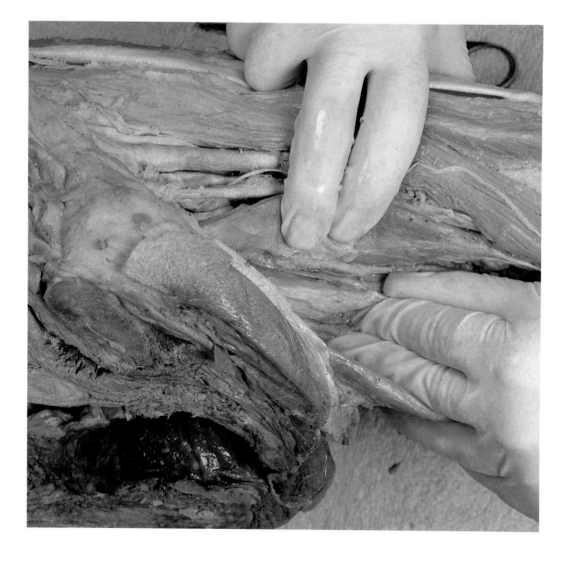

The Gluteal Region

In the gluteal region, the sciatic nerve is the most important structure, with the piriformis muscle being an important landmark as the nerves and vessels entering the gluteal region are closely related to it. Both will be studied in detail, and the gluteal muscles, vessels and nerves will then be demonstrated. The branches of the lumbosacral plexus are traced from the pelvis to the gluteal region. The path of the pudendal nerve is followed from the pelvis, through the gluteal region, to the perineum.

Make a transverse incision in the skin on the back of the thing, approximately two inches below on the gluteal fold.

Remove the fatty superficial fascia from the buttock. Identify any cutaneous nerves that you discover.

292 A Define the borders of the *gluteus maximus muscle*. Use your fingers to separate this muscle from the underlying structures; define its origins and insertions **(1)**.

1

*Before proceeding with the dissection, define the position of the **lower border of the piriformis muscle** in your specimen:*
- *find the **posterior superior iliac spine**,*
- *find the **tip of the coccyx**,*
- *find the mid-point of a line joining these two structures,*
- *a line from this point to the top of the **greater trochanter** corresponds to the lower border of the piriformis muscle.*

Make a small incision in the *gluteus maximus* which corresponds with the midpoint of the lower border of the piriformis muscle. Enlarge this incision by separating the fibres of the muscles with your fingers.

Feel through this aperture for the sciatic nerve, using your thumb and forefinger. Take careful note of the position of this nerve, with respect to the gluteus maximus muscle and to the eminence and fold of the buttock.

Cut through the gluteus maximus muscle, two or three centimetres from its origin, and reflect it laterally. Identify and clean the *piriformis muscle*. Clean and demonstrate the structures shown in the illustration.

*Identify the **gluteus medius muscle** and the nerves and vessels around the piriformis muscle.*   **293 B,C**

Divide the gluteus medius muscle near to its insertion into the greater trochanter, and reflect it upwards.

*Identify the gluteus minimus muscle and the **superior gluteal nerve and vessels**.*
*Turn the specimen over and identify the **psoas major** **and iliacus muscles** in the posterior abdominal wall.*   **244 A**
*Use your fingers to trace these muscles to their insertions on the **lesser trochanter of the femur**.*   **243 B**
*Identify the **branches of the lumbar plexus**, and trace the **lateral femoral cutaneous nerve**, the **obturator nerve** and the **femoral nerve** into the lower limb.*

Remove the *psoas muscle* piecemeal, to demonstrate the structure of the lumbar plexus. Divide the *piriformis muscle* close to its origin, and reflect it laterally to observe the *sacral plexus*. Trace its branches through the *greater sciatic foramen*, into the gluteal region.

*Find the **greater and lesser sciatic foramina** in the gluteal region by palpation.*
Follow the course of the sciatic nerve from the plexus to the gluteal region.
*Follow the **pudendal nerve** around the **ischial spine** into the perineum.*

The Back of the Thigh and the Popliteal Fossa

In this dissection the path of the hamstrings and sciatic nerve is followed. The popliteal fossa is an important topographical region located at the junction between the thigh and the leg.

Remove the skin from the back of the thigh and discard the flap. Make a further incision on the medial side of the leg, and reflect a skin flap laterally to the mid-calf level.

Remove the superficial fascia from the back of the thigh and upper part of the leg.

> *Identify any **cutaneous nerves** that you uncover, and preserve the upper end of the **small saphenous vein**.*

Make a longitudinal incision in the deep fascia from the gluteal region to the calf; reflect the flaps medially and laterally, taking care not to damage the muscles forming the boundaries of the popliteal fossa.

Remove the connective tissue from the popliteal fossa, and define the contents of the fossa (**1**).

1

294 B
306 A,B
307 C

Identify the borders of the popliteal fossa, as well as the nerves and vessels passing through.

> *Separate and identify the **hamstring muscles**, from their origins at the ischial tuberosity and upper part of the femur, to their insertions (**2**). Trace the **sciatic nerve** through the posterior compartment of the thigh.*

295 C

2

> *Pull the **biceps femoris** medially to examine the posterior surface of the **adductor magnus muscle**, and identify the branches of the **profunda femoris artery** which supply the structures in the posterior compartment of the thigh.*

> *Turn the specimen over and observe the continuity of these branches with the **profunda femoris artery** in the anterior compartment of the thigh. Show that the **femoral artery** enters the popliteal fossa through the opening in the adductor magnus muscle, to become the **popliteal artery**.*

The Front of the Leg and Dorsum of the Foot

The leg contains three osteofascial compartments. In this dissection the anterior and lateral compartments are studied, and the tendons of muscles in the anterior compartment are followed on to the dorsum of the foot.

Make an incision on the front of the leg and dorsum of the foot, reaching the roots of the toes. Make transverse incisions below the knee, between the malleoli, and at the roots of the toes. Take care, as the skin is thin and there is little superficial fascia on the dorsum of the foot. Reflect the skin flaps laterally.

Dissect and clean the *great saphenous vein* medially and *small saphenous vein* laterally on the dorsum of the foot (**1**).

1

Use blunt dissection to remove the fat and most of the superficial veins from the anterior surface of the leg. Preserve the saphenous veins and nerves.

*Identify the **deep fascia** of the leg and **superior extensor retinaculum**.*

*Identify the **anterior (extensor)** and **lateral (peroneal) compartments** through the deep fascia.*

Make a longitudinal incision in the overlying deep fascia to open the anterior compartment; be careful to retain the superior extensor retinaculum.

*Separate and identify the following muscles (**2**) :*
○ *tibialis anterior,*
○ *extensor hallucis longus,*
○ *extensor digitorum longus,*
○ *peroneus tertius.*

Trace their tendons to the superior extensor retinaculum and dorsum of the foot.

314 A,B

2

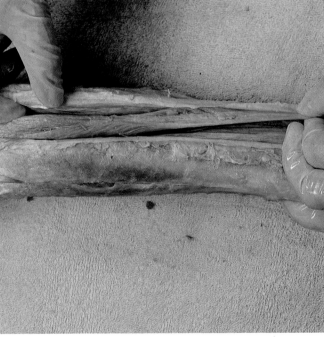

*Trace the great and small saphenous veins into the **dorsal venous arch** lying very superficially in the foot.*

*Find the **saphenous nerve** which lies close to the great saphenous vein and trace it into the leg.*

*Find the **superficial peroneal nerve** on the surface of the deep fascia in the lower third of the leg, and trace both nerves into the dorsum of the foot. Try to locate the branches of the **lateral cutaneous nerve of the calf**.*

Make a longitudinal incision and reflect the deep fascia to open the lateral peroneal compartment. Be careful to avoid damaging the superior extensor retinaculum .

*Separate and identify the **peroneus longus and peroneus brevis muscles**, but do not trace the tendons into the foot at this stage (3).*

*Note the **intermuscular septum** extending from the deep fascia and separating the lateral and anterior compartments of the leg.*

68

Dissection of the Back of the Leg and relations of the Ankle Joint

In this dissection the great and small saphenous veins are traced further, and the superficial muscles of the posterior compartment of the leg are demonstrated.

Then, the deep contents of the posterior compartment of the leg are demonstrated. The nerves, vessels and tendons passing from the leg into the foot from all three compartments are subsequently examined as relations of the ankle joint.

Superficial dissection

Remove the skin from the back of the leg to the heel and over the sides of the foot, taking great care to avoid damaging the superficial structures.

Clean the *great and small saphenous veins* to follow their full course. Find the *sural nerve* beside the small saphenous vein.

Revise the borders and contents of the popliteal fossa.

Make a longitudinal incision in the deep fascia; reflect the flaps medially and laterally, taking care not to damage the muscles which form the boundaries of the popliteal fossa.

12 A,B
313 C

*Examine the superficial muscles of the posterior compartment, and separate them with your fingers. The two heads of the **gastrocnemius muscle** may be separated by pulling them apart to expose the **plantaris and soleus muscles**. Trace the latter to their origins and insertions.*

Deep dissection

Divide the *tendo calcaneus* (Achilles tendon) and reflect the gastrocnemius upwards (**1**). Cut through the fleshy origin of the soleus on the tibia; reflect the muscle to expose the *deep transverse fascia* between the superficial and deep muscles of the posterior compartment.

Make a longitudinal incision in the deep transverse fascia.

1

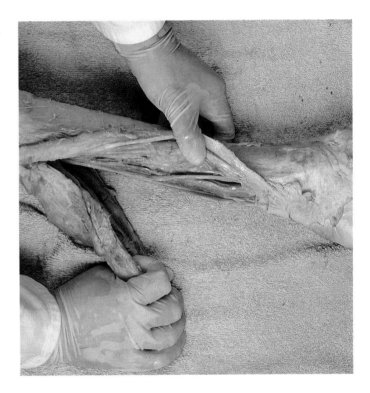

Identify and separate with your fingers the deep flexor muscles of the posterior compartment of the leg. Trace their tendons downwards, but not all the way into the sole of the foot.

315 C,D
316 A
317 C

*Identify the **neurovascular bundle** of the posterior compartment, and trace its contents from the popliteal fossa to the flexor retinaculum. Divide the flexor retinaculum, but leave its cut edges in place.*

*Determine the relations between the tendons, vessels and nerves as they pass behind the medial malleolus (**2**). Trace these structures back into the leg.*

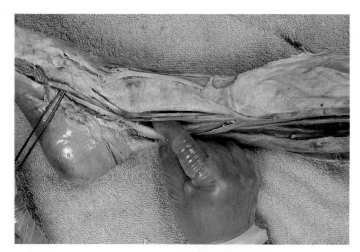

2

Identify the **superior extensor retinaculum** and **inferior extensor retinaculum**. Cut through the inferior extensor retinaculum, but preserve its attachments. Trace the following tendons to their insertions (**3**):
○ **extensor digitorum longus**,
○ **extensor hallucis longus**,
○ **tibialis anterior**.

Identify the **extensor digitorum brevis** muscle and trace its four tendons to their insertions (**3**).

Examine the insertions of extensor muscles in the second or third toe. Define the **extensor expansion** over the proximal phalanx, and trace its branches to the middle and distal phalanges.

Identify the tendons of the small **lumbrical** and **interosseous muscles** of the foot.

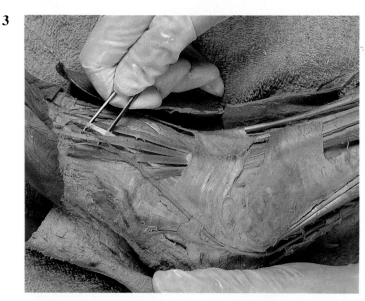

3

Return to the anterior compartment of the leg. Identify and clean the **anterior tibial artery**, its companion veins and deep peroneal nerve, lying on the **interosseous membrane**.

Similarly, find the **common peroneal nerve** at the superior end of the lateral compartment. The superficial peroneal nerve lies within the peroneus longus muscle until it reaches the level of peroneus brevis. It then passes anteriorly and runs between the peroneus brevis and extensor digitorum muscles, piercing the deep fascia at the junction between the middle and distal thirds of the leg.

Trace all these vessels and nerves upwards and downwards. Make a longitudinal incision in one of the larger veins to observe the valves.

Find the **dorsalis pedis artery** on the dorsum of the foot by pushing aside the tendons of extensor digitorum longus and extensor hallucis longus. Find the **deep peroneal nerve** lying beside it. Trace the branches of this artery and nerve to the toes, and confirm the continuity of these structures with the anterior tibial compartment of the leg.

Find the **perforating branch of the peroneal artery** in the anterior compartment of the leg, deep to the **peroneus tertius muscle**, and trace this artery into the foot

Identify the **peroneal retinaculum**, and the tendons passing behind the lateral malleolus (**4**).

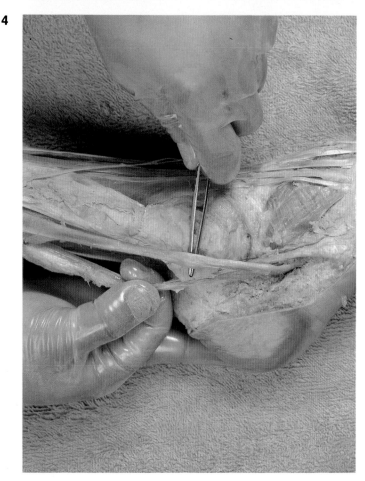

4

319 C, D

69

The Sole of the Foot

This is a difficult dissection. You will have to display delicate structures running through dense connective tissue. Use a sharp scalpel and fine forceps, and be careful to clean each structure as it is uncovered in the course of dissection.

Remove the skin and superficial fascia from the sole of the foot and flexor surfaces of the toes, by starting at the heel and reflecting both layers forward to the toes (**1**). Avoid damaging the *plantar aponeurosis*.

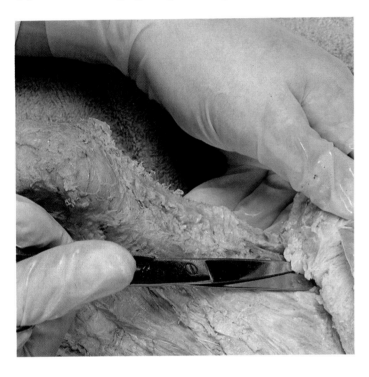

324 A   Clean the *plantar aponeurosis*.

Remove the plantar aponeurosis piecemeal to display the first layer of muscles of the sole (**2**).

325 B
Identify:
○ *the **adductor hallucis**,*
○ *the **flexor digitorum brevis**,*
○ *the **adductor digiti minimi**.*
*Search for the **medial and lateral plantar nerves and vessels**. Establish the continuity of these nerves and arteries with the **tibial nerve** and **posterior tibial artery**, behind the medial malleolus and in the posterior compartment of the leg.*

Separate the flexor digitorum brevis muscle from the underlying structures. Remove the belly of the muscle, but leave the tendons attached to the toes. Clean and identify the muscles of the second layer of the sole.

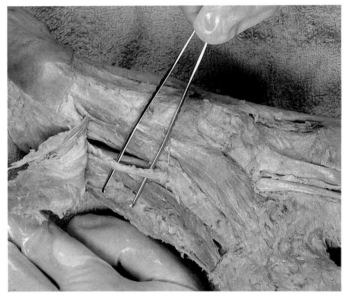

*Identify the **tendon of flexor digitorum longus**, **flexor accessorius** and the **lumbricals**. Note that the tendon of flexor hallucis longus crosses the sole deep to the tendon of flexor digitorum longus. Identify both long flexors in the posterior compartment of the leg.*   **325 C**

Divide the tendon of flexor digitorum longus just where it is joined by the flexor accessorius. Reflect the anterior part forwards, to display muscles of the third layer of the sole in the anterior half of the foot. Clean these structures.

*Identify the **flexor hallucis brevis**, **flexor digiti minimi** and the two parts of **adductor hallucis**. Cut through the origin of the flexor hallucis brevis, and reflect the muscle forwards to display the two **sesamoid bones** in its tendons below the head of the first metatarsal.*

*Divide the oblique head of adductor hallucis, and reflect the muscle to display the **plantar arterial arch** and **deep branch of the lateral plantar nerve**. Clean the plantar arterial arch and determine its formation and branches.*   **326 C**

*The fourth layer consists of the **interossei**. Force the metatarsals apart, to observe their attachments.*   **326 B**

Remove the second or third toe by forcing the metatarsals apart, and cutting through the soft tissue in the intermetatarsal joint. Use forceps and a seeker to separate the tendons and muscles still attached to the toe.

Identify these structures and define their insertions.

Examine the ligaments of the sole of the foot: divide the flexor accessorius at its origin, and turn it to one side to display the *long plantar ligament* and *plantar calcaneonavicular (spring) ligament.*

*Explore the osteofibrous tunnel of the **peroneus longus tendon**.*   **327 D**

*Find the tendons of the **peroneus brevis** and **tibialis posterior muscles**, and define their insertions.*

Trace all three muscles back into the leg.

Divide the long plantar ligament to display the *short plantar ligament* and insertion of the *peroneus longus tendon.*   **327 E**

70

The Hip Joint

*For this dissection choose a limb which will be least useful for revision of
the muscles, nerves and vessels, as the structures surrounding the joint
must be removed before the joint capsule is observed and opened.*

1

2
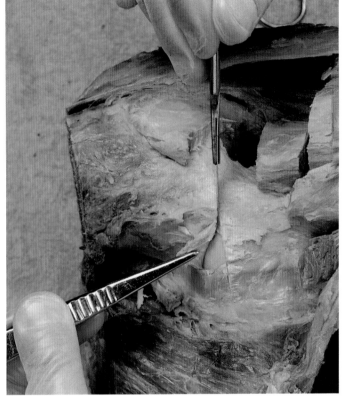

*Before starting to dissect, study carefully the hip
bone, femur and the movements at the joint in the
living subject.*

*Return to your dissection to identify all the muscles
related to the hip joint.*

Make a transverse incision through the *iliofemoral
ligament* (**2**). Note its thickness and strength. Continue
the incision posteriorly through the other ligaments
and capsule of the joint. Dislocate the joint by pulling
on the femur.

*Examine the articular surfaces of the acetabulum and
head of the femur.*

301 D

*Examine the attachments of the capsule of the joint to
the femur and hip bone.*

00 A,B Cut through each muscle at its insertion, and divide the
muscle bellies more proximally to expose the ligaments
of the joints (**1**).

71

The Knee Joint

Use the same limb as in the dissection of the hip joint. The functions of the ligaments of the knee joint will be studied, as well as the articular surfaces and menisci.

Study the lower end of the femur, upper end of the tibia and patella; pay particular attention to their articular surfaces.

Identify all the muscles related to the knee joint on your specimen.

Divide the superior muscles just above their tendons, and the two heads of the *gastrocnemius* just distally to their origins on the femur.

307 D Remove the vessels and nerves from the *popliteal fossa,* and clean the posterior surface of the *capsule of the knee joint* and *popliteus muscle.*

308, 309 Clean the *tibial* and *fibular collateral ligaments,* and the *oblique ligament* of the knee joint. Define the origins and insertion of the *popliteus muscle.*

Divide the *patellar retinacula* and *quadriceps tendon;* turn the *patella* and *patellar ligaments* downwards.

Observe the tibial and fibular collateral ligaments, oblique ligament and popliteus muscle (1 & 2).

*Explore the **suprapatellar** and **infrapatellar bursae.***

1

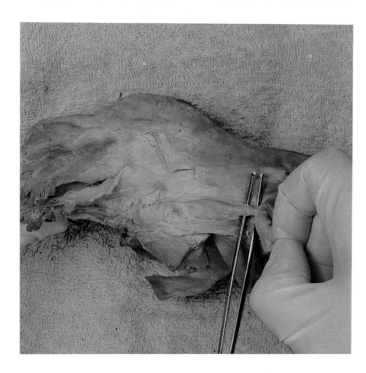

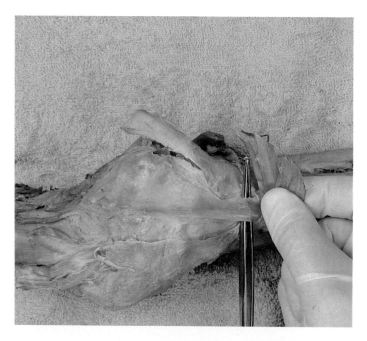

2

Cut through the lateral and medial parts of the *capsule* of the knee joint, as far as the *tibial* and *fibular collateral ligaments.*

Flex the joint and observe the intra-articular structures:
- *the **transverse ligament,***
- *the **infrapatellar synovial fold** and **alar folds,***
- *the **menisci,***
- *the **cruciate ligaments.***

**310 A
308 A,
B,C**

Test the stability of the joint with the ligaments intact. To appreciate the function of each ligament in the stability of the joint, divide the ligaments one at a time, and test the stability of the joint between each division.

Divide the ligaments in the following order:

1. Fibular collateral ligament.
2. Tibial collateral ligament.
3. Anterior cruciate ligament.
4. Posterior cruciate ligament.

Separate the *femur* from the *tibia.*

*Study the articular surfaces of the **femur, patella** and **tibia,** the attachments of the infra-articular structures.*

309 E

72

The Ankle and Subtalar Joints

In this dissection the ankle and subtalar joints are examined. The ankle (talocrural) joint, between the tibia and fibula and the talus, allows plantar flexion and dorsiflexion to take place. The movements of inversion and eversion occur at the subtalar and talocalcaneonavicular joints.

Remove all muscles and tendons around the ankle joint (**1**).

1

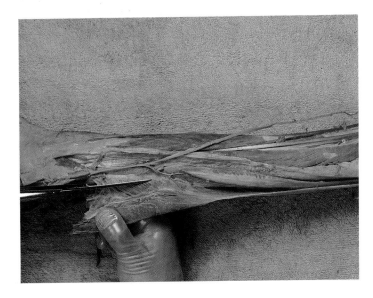

Take note of their origins and insertions as you do so.

323 C,D Clean the ligaments and capsule of the ankle joint (**2**).

2

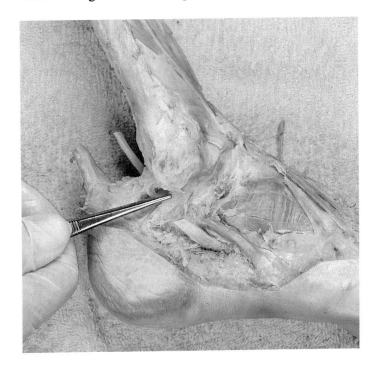

Identify and note the thickness, and hence the strength, of the ligaments on the lateral and medial sides of the joint.

Cut through the anterior, lateral and posterior aspects of the joint capsule and ligaments. The leg and foot are left attached by the medial ligament (**3**).

3

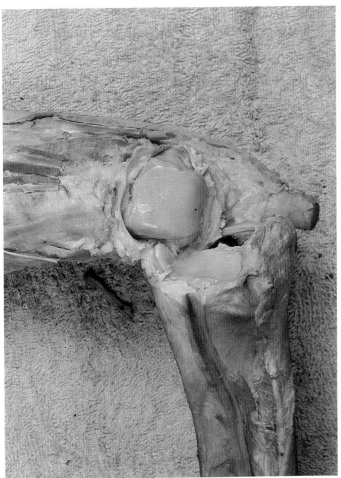

Examine the articular surfaces of the ankle (talocrural) joint.

Cut through the medial (*deltoid*) ligament, to remove the leg from the foot.

Display the subtalar joint by removing the talus. In order to do this, the bone, and therefore the capsule of the joint, should be defined. Use scissors to cut through the capsule (**4**).

Use scissors to divide the interosseus ligament (**5**). Remove the talus.

4

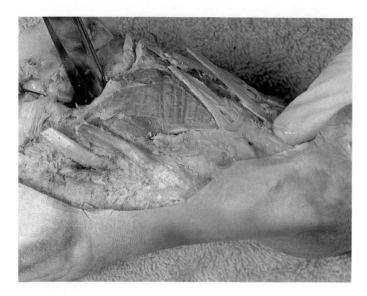

5

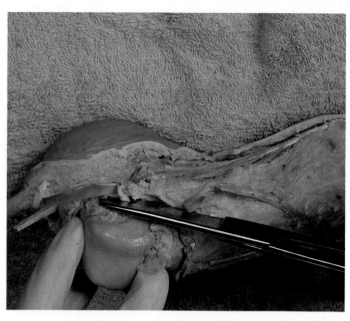

Note that the talus is still attached to the foot by the **interosseous (subtalar) ligament**.

Examine the articular surfaces of the talus, calcaneum and navicular bone.

322 B

73

Osteology of the Lower Limb

63-286

Use the illustrations in the atlas to identify all the major parts and markings on:
○ *the **hip bone**,*
○ *the **femur**,*
○ *the **patella**,*
○ *the **tibia**,*
○ *the **fibula**.*

Note the areas on the posterior surface of the patella that articulate with the medial and lateral femoral condyles.

Note the differences in shape and location between the medial and lateral condyles of the tibia.

Using the illustrations in the atlas, review all the muscles and ligamentous attachments to the bones listed above.

Place the bones in their correct anatomical position, and determine the side of the body to which each bone belongs.

Map out the three parts of the hip bone:
○ *the **pubis**,*
○ *the **ischium**,*
○ *the **ilium**.*

*On an articulated skeleton, note the following features of the **femur** and **hip joint**:*
○ *the direction of the neck of the femur,*
○ *the angle between the neck and body of the femur,*
○ *the line between the centre of the femoral head and the centre of the knee joint (this is the line, or axis, around which the thigh rotates),*
○ *the 'bowing' of the femoral shaft; this is convex anteriorly.*

Identify the parts of the bones contributing to the following joints:

● *The **subtalar joint**, between a facet on the inferior surface of the body of the talus, and a posterior facet on the calcaneus.*

● *The **transverse tarsal joint**, being two separate joints:*
○ *the **talcalcaneonavicular joint**, between the head of the talus and (a) the anterior and middle facets of the calcaneus;*
(b) *the articular cartilage of the calcaneonavicular ligament;*
(c) *the posterior articular surface of the navicular bone.*
○ *the **calcaneocuboid joint**, between articular facets on the calcaneus and cuboid bones.*

87-289

Using the atlas, identify the bones of the foot. In addition, try to identify them on as many radiographs of the foot as possible.

Joints of the foot
These are complex joints; to understand them, carry out the following exercise:

● *Obtain an articulated set of bones of the foot, joined by strings. **Do not use an articulated skeleton in which the bones are held together by wires.***

● *Pull the strings fairly tight and hold the talus in one hand, holding the rest of the foot in the other hand.*

● *Reproduce the movements of **inversion** and **eversion**.*

74

Surface Anatomy of the Lower Limb

In this section the bony landmarks, nerves and blood vessels of the lower limb will be studied first; the contours and muscles of the limb, as well as the basic movements at the joints will then be observed. Finally, the lower limb will be studied in action.

On a living subject, other than yourself, demonstrate the following bony landmarks:

In the hip and gluteal regions, find:
○ the **highest point on the iliac crest,**
○ the **anterior superior iliac spine,**
○ the **posterior superior iliac spine,**
○ the **ischial tuberosity,**
○ the **pubic tubercle,**
○ the **greater trochanter of the femur.**

In the region of the knee, find:
○ the **patella,**
○ the **line of the knee joint,**
○ the **medial** and **lateral condyles** and **epicondyles** of the femur,
○ the **tibial tuberosity,**
○ the **head and neck of the fibula.**

In the lower part of the leg and foot, find:
○ the **subcutaneous parts of the tibia and fibula,**
○ the **medial malleolus,**
○ the **sustentaculum tali,**
○ the **tuberosity of the navicular bone,**
○ the **head of the talus,**
○ the **lateral malleolus,**
○ the **peroneal trochlea,**
○ the **bases and heads of the metatarsals,** the intertarsal spaces,
○ the **tuberosity on the base of the fifth metatarsal,**
○ the **cuneiform bones.**

Nerves
Review the surface markings of the sciatic nerve in the gluteal region (see Section 65).

Roll the common peroneal nerve on the neck of the fibula.

Blood vessels
The pulse of the **femoral artery** can be felt just below the **midinguinal point,** halfway between the pubic symphysis and the anterior superior iliac spine.

The pulse of the **popliteal artery** can be found in the popliteal fossa. The subject should flex the knee to relax the fascia in the popliteal fossa, and the artery should then be pressed against the back of the tibia.

The **dorsalis pedis artery** can be felt between the tendons of extensor hallucis longus and extensor digitorum longus, in front of the ankle joint.

The pulse of the **posterior tibial artery** can be felt posterior to the medial malleolus.

Superficial veins
Note the origins of the **great** and **small saphenous veins** around the ankle.
The **great saphenous vein** can be found on the medial side, one hand's breadth behind the patella. It ends at the saphenous opening, 1.5 inches below the **groove of the groin,** just medial to the femoral artery.

Muscles
Identify the **anterior, medial** and **posterior compartments** of the thigh, and name the muscles in each compartment.

Identify the **anterior, posterior** and **lateral compartments** of the leg. For the lateral compartment, ask the subject to press down on the medial side of the foot. List the muscles in each compartment of the leg.

Identify the muscles and tendons forming the boundaries of the popliteal fossa.

Movements
At the **hip joint,** demonstrate:
○ flexion,
○ extension,
○ abduction,
○ adduction.
With the subject's knee flexed, demonstrate **rotation** at the hip joint.

At the **knee joint,** demonstrate:
○ flexion,
○ extension.

With the subject in the sitting position, demonstrate rotation of the tibia on the femur.

At the ankle joint, demonstrate:
○ *dorsiflexion,*
○ *plantar flexion.*
Demonstrate inversion and eversion, with the subject's foot:
(a) fully dorsified,
(b) in a neutral position,
(c) fully plantar-flexed.

Functional Anatomy

Standing
Ask the subject to stand upright with the feet slightly apart. Determine which groups of muscles are active. Note any changes in muscle activity as the subject sways backwards and forwards. Ask the subject to stand on one leg. Note any changes in posture at the hip and lumbar spine, as well as changes in muscle activity. Try to make the same observations on yourself.

Standing up from a seated position
Watch the subject as he or she stands up from a comfortable sitting position. Note the movements at the hip, knee and ankle, and determine which muscles are active. Repeat these observations on yourself.

Walking
Starting to walk: when the subject starts to walk forwards, what is the first movement involved? Which muscles are responsible?

Walking forwards: carefully observe the subject from the side, as he or she walks slowly forwards. Record the movements occurring in the form of othe gait cycle. Observe and record the movements occurring at the hip, knee and ankle joints, first for the supporting limb and then for the swinging limb.

	Support Phase	Swing Phase
Hip		
Knee		
Ankle		

Observe the subject from the front, and note the tilting of the pelvis during walking. Try to observe the rotation of the pelvis around a vertical axis, as the subject walks forwards.

Now observe the subject's foot very closely, during the stance phase of walking. This phase is often subdivided as follows:

Heel Strike	Mid-Stance	Heel Off	Toe Off

Note what happens at the ankle and digits, at each of these stages. Walk forwards yourself, and try to visualize what is happening in your own foot. Note how, and when, your weight is carried by the various parts of the foot during the support phase, and how it is transferred to the other foot at the start of the swing phase. Correlate this weight transfer with the rotation of the pelvis.

Appendix

Alternative Approaches

The scheme adopted in this book has the advantage of treating the most complicated part of the body first, but it is possible to start with any of the other regions.
The osteology and surface anatomy are placed at the end of each region, but may be studied before beginning the dissection.

Upper Limb

Start with Section 21, but omit the paragraph in Section 26 (The Back) where the upper limb is detached from the body, so that the clavicle remains in place for the dissection of the neck.

Thorax

Start with Sections 23 and 24 of the upper limb dissection, so that the thoracic wall is accessible. It will be impossible to make the saw cut to remove the anterior wall of the thorax as far posteriorly as the instructions suggest, and it will be difficult to see the structures in the superior mediastinum until the head and neck have been fully dissected.

Abdomen
Lower Limb

The dissection may start with either of these regions, without any change in the scheme.

Comments

Head and Neck

The scheme adopted in this book is to make a superficial dissection of the neck and head, and then to work from the cranial cavity outwards. This approach provides a clear understanding of the anatomy of the pharynx and the carotid sheath and its contents; these are structures which are often not adequately treated in other dissection approaches, and consequently not fully understood by students.
The study of the larynx has been distributed in Sections 10, 11, 12 and 15.

Upper Limb

The back (Section 26) has been dissected with the upper limb. This is because it is convenient to continue with the dissection of the back where the upper limb is detached from the body.

Perineum

There is some advantage in dissecting the perineum before the pelvis is bisected, although it is sometimes difficult to abduct the lower limbs sufficiently for a proper dissection of this region. If the lower limbs can be abducted, then the perineum could be dissected after the anterior abdominal wall and inguinal region.

Further Reading

There are many textbooks of anatomy available, and as the differences between them are largely matters of style, choose a book that suits you. As anatomy can be a daunting subject, a general textbook should not be too long; about 500 pages is adequate.

Hollinshead, W.H.: Anatomy for Surgeons (3 vols), second edition, Harper & Row, 1982.

Kapandji, L.A.: The Physiology of the Joints (3 vols), second edition. Churchill Livingstone, 1974.

Pegington, J.: Clinical Anatomy in Action (3 vols), third edition. Churchill Livingstone, 1988.

In addition, refer to all the relevant current atlases by Wolfe Publishing:

Backhouse, K.M. and Hutchings, R.T.: A Colour Atlas of Surface Anatomy (1986).

England, M.A. and Wakely, J.: A Colour Atlas of the Brain and Spinal Cord (1988).

McMinn, R.M.H., Hutchings, R.T. and Logan, B.M.: A Colour Atlas of Head and Neck Anatomy (4th impression 1985).

McMinn, R.M.H., Hutchings, R.T. and Logan, B.M.: Picture Tests in Human Anatomy (1986).

McMinn, R.M.H., Hutchings, R.T. and Logan, B.M.: The Human Skeleton, a Photographic Manual (1986).

McMinn, R.M.H., Hutchings, R.T. and Logan, B.M.: A Colour Atlas of Foot and Ankle Anatomy (1982).

Wirhed, R.: Athletic Ability and the Anatomy of Motion (1984).

INDEX